SELLING WITH COLOR

This book is produced in full compliance with the government's regulations for conserving paper and other essential materials.

SELLING WITH COLOR

by
FABER BIRREN

First Edition
Third Impression

New York London
McGRAW-HILL BOOK COMPANY, INC.
1945

SELLING WITH COLOR

PRINTED IN THE UNITED STATES OF AMERICA

THE MAPLE PRESS COMPANY, YORK, PA.

To

HERBERT KAUFMAN

PREFACE

THIS is a book about color and people. It has been written to be of practical benefit to modern business, to assure the effective development of consumer products, merchandising, advertising, packages, displays.

Most books on color will be found long on theory and short on practice. This, perhaps, is because color is emotional in its appeal and tends to inspire a personal and subjective viewpoint. Yet while the so-called artistic aspects of color may be elusive, the facts of mass human reaction are otherwise. For color in industry today may be "engineered" with remarkable certainty, once the methods are known and applied with intelligence and care.

It is evident that when you pick the *right* colors you sell a lot of merchandise or influence a lot of people; when you pick the *wrong* colors you pile up unsold inventories and see the public turn its back.

To management this means that color problems must be rightly analyzed and estimated, that customers and markets must be known.

To those charged with the details of styling, the creative "hunch" must be supported by a real understanding of human wants and desires.

It is the intent of this book to present facts rather than opinions and to set forth principles that have the support of extensive research and sales record. To accomplish this it has been necessary to gather all possible data from the scientific investigator and to study the sales experience of numerous industries. With such evidence at hand the more temperamental aspects of color may be thrown into clearer light and a better and surer control realized.

American business knows the value of color. Yet it also is aware of certain hazards and risks. If color is to build profits, if those who use it are to find success in their efforts, a more accurate knowledge of the public heart and mind must be achieved.

A good bit of such knowledge will, it is hoped, be found in the pages of this book.

FABER BIRREN.

NEW YORK,
March, 1945.

CONTENTS

PAGE

Preface vii

CHAPTER 1

Public Taste—What Is It? 1

CHAPTER 2

These Are the Colors That People Prefer 16

CHAPTER 3

Science Offers an Answer 29

CHAPTER 4

These Are the Things That People Buy 42

CHAPTER 5

Giving the Public What It Wants 49

CHAPTER 6

Practical Research Techniques 63

CHAPTER 7

Glorifying Human Desires 74

CHAPTER 8

The Human Nature of Vision 88

CHAPTER 9

More Power to Advertising 104

Page

CHAPTER 10
Packages, Displays, Interiors 116

CHAPTER 11
The Art of Color Conditioning 128

CHAPTER 12
Color Enterprise—Plain and Fancy 140

CHAPTER 13
The Psychology of Color 159

CHAPTER 14
The Romance of Color 173

CHAPTER 15
The Specification of Color 190

APPENDIX A
A Review of Sales Records 209

APPENDIX B
A Check List of Color Standards 222

Annotated Bibliography 229

Index 235

Chapter 1

PUBLIC TASTE—WHAT IS IT?

A GOOD many years ago Henry Ford remarked that he wouldn't give a dollar for all the art in the world. Those with a flair for the aesthetic and the cultural may consider these words harsh indeed. Coming from a capable industrialist, however, they hold practical elements of good sense. After all, the problem of business is to sell merchandise or otherwise to influence customers and markets. To take the attitude that business ought at the same time to elevate the artistic discrimination of its public is beside the point, for anything more unstable and fugitive than good taste would be hard to find.

Business has one great virtue in this regard—it tries its best to satisfy people. If many of its products are deemed ungainly by some current standard, they are at least salable or they would not be made. The artist or the designer who argues that such merchandise is bad may speak outside his province. The businessman wants to know what he is about. For him to say that "beautiful merchandise is that which sells" is not poor philosophy, nor is it really deserving of rebuke.

No doubt, the salt-box house, the homespun garments, the furniture, rugs, and textiles of the New England Yankee were in definitely bad taste in the judgment of those Tories who were lucky enough to be more opulent and sophisticated. Such more or less primitive ideas of art conceivably brought forth smiles from noble ladies and gentlemen who had promenaded through the lofty, paneled halls of English mansions. Yet, as the centuries passed, the simple Yankee has proved to have had truly good taste, for the things he loved and cherished have survived. Possibly, the loving and cherishing helped to endow them with an unmistakable human beauty, to be appreciated by other generations.

However, there is hardly much need for romancing in this book on color and people. The facts are evident: if you would satisfy some high individualism with color, realize that your pleasure will not necessarily be the pleasure of others. On the other hand, if you would style merchandise for mass markets, decorate buildings to attract crowds, design advertising and displays to appeal to millions, then you must submerge individual, personal notions in the bigger sea of human life. You must think objectively and not subjectively.

The difference is this: color in the fine arts gives expression to the creative spirit of the individual artist. Color in industry, inevitably more democratic, attempts to comprehend and to satisfy the desires of the public at large.

Markets and Colors

As a first practical consideration, there are two distinct markets for color, to be served through two distinct methods. Call one high fashion and the other mass.

In the high-fashion market there is more of a show of wealth, the difference between $1.98 and $2 being of minor consequence. The people here are discriminating, show excellent taste, and strive for an individuality that they seldom, unfortunately, achieve. In dress they have ephemeral likes and dislikes, worrying more about the cut of a material than about its wearing qualities. To be different is something of a credo with them—although, in the attempt to achieve this, they may all look strangely alike.

They are extremely conscious of color coordination and harmony. Using the head rather than the heart, they will buy apparel more for the sake of vogue than for the appeal color may hold upon their emotions. In consequence, they are individual style leaders and good prospects for the manufacturer who would rather sell few things at a large unit profit than many things at a small margin.

In home decoration, the high-fashion market is one that follows advanced trends. Everything purchased must fit in with a well-ordered ensemble. Products must match or at least blend concordantly. Choice here is dictated not by emotional impulses, but rather by conventions established by magazines, interior decorators, department stores, and makers of more costly merchandise. High-fashion buyers are literate, well tutored in what goes on in a changing world, and sufficiently financed to ride high on new tides.

While this market is small in size, it is a profitable one and will pay a good price for things exceptional. From the manufacturer's standpoint, it is a market of speculation and risk, for it demands novelty, originality, and untried ventures. As far as color is concerned, here are the principles that seem to apply.

1. Color effects must avoid the commonplace.

2. Off shades will appeal if they are carefully coordinated with related products.

3. Individual colors must *fit in* with well-conceived notions of ensembled wardrobes and ensembled interiors.

4. The styling effort must contemplate the whole picture rather than its single parts. This means that the manufacturer has a big job to tackle in keeping abreast of color trends in many lines of merchandise.

5. As swiftly as any one idea becomes acceptable (and thereupon grabbed for mass markets), the new thing must be offered to overcome the ennui of those who presume to lead the community and who have the cash to do so.

6. Trends in high-fashion markets move at a fairly rapid rate. It is important to watch shifts in preference—both to capitalize rising demands and to avoid leftover inventories. Wide color ranges are desirable, to cater to the vagaries of sophisticated taste.

But a great number of high-fashion buyers are graduates from mass thinking, just as adults are the enlargement of children. Remember that the schoolgirl who once insisted upon dressing exactly like her classmates may one day become the woman who will grow pale at seeing the duplicate of her dress or hat in a cocktail lounge. Yet the mass taste may rest quite easy with her in things less personal than an evening gown or a suite of parlor furniture—a fountain pen, for example, or a Mickey Mouse doll, which may catch her as the eternal child.

However, it is the mass market that is of major importance to the average modern business (and that forms the chief interest of this book). How are millions to be satisfied with color? What are the patterns of their desires as they crowd the stores and catalogues of the nation? What is the magic that rockets one color to incredible heights of volume, while another staggers and falls by the way?

Consider now the mass buyer as distinct from the high-fashion buyer. Her budget is lean. There are matters more pressing than taste. She will, perhaps, look at the price first, examine the quality second, and then grant third place to her emotional feeling. However, let this emotional feeling be compelling enough, and she will fling open her purse with almost reckless abandon.

Having budgets to think about, the mass buyer is not able to cater to whims or to be too fussy about ensembled wardrobes or rooms. Every investment with her is a sizable one, and quite naturally everything she buys must have a conspicuous beauty! *It must stand out by itself long before it must fit in with anything else!* How otherwise would the neighbors know that she had purchased something new?

Understand this characteristic in human beings—one that is really social and economic in its basis—and you will better appreciate the essential difference between the high-fashion and the mass-market buyer, high income and low income.

In apparel, the normal prospective wearer will weigh colors, style, and material against the practical question of wearing quality, with simple elemental colors in predominance. In home furnishings, she will readily accept gaudy hues that all bespeak frank delight to the senses, whether or not they will fit together.

Thus, for a second set of principles, compare this with the one previously given.

1. In low-income markets the color range must be simple, compelling, and rich or bright.

2. Off shades, strange and subtle colors are not wanted.

3. Individual hues and hue effects must stand out by themselves in unblushing glory.

4. The styling effort seldom has to contemplate coordination with other merchandise. While it is perhaps desirable,

coordination should never be attempted merely for its own sake if it means the compromise of frank, clean colors.

5. Past sales records are more important than are style prognostications. Mass buyers are followers, not leaders.

6. Style trends are slow in movement. Low-income buyers seldom weary of the colors they like best. Their fancies, having an almost instinctive basis, are not subject to radical changes. Restricted color ranges are therefore best, and these should be carefully checked, through research.

Design and Color

It may be said that in the styling of consumer merchandise, advertising, or displays, the element of design seems to stimulate mental processes in the buyer, while the response to color is largely emotional. While this reaction is not invariable, of course, it is typical of the attitude that most persons assume in spending their money or in being swayed to do so. To understand this is to know why color is a more sensitive thing to handle than is shape, or form.

The streamlining of an automobile has completely changed its aspect in the period of a few years. These improvements have been praised and accepted. Is it that men and women crave newness? If so, then the question should be limited to design. For the attempt to do equally radical things with color has persistently failed. Form is something for the brain to appreciate, and the brain may be educated. Color, however, is more like religion. It is in the blood, an essential part of the psychic make-up of an individual. The buyer may relish the new fender lines and radiator grille but, thank you, he will prefer the color to be black or maroon, blue or green. Surely if you give him an automobile the likes of which he hasn't seen before, then he ought to complement his purchase with an equally unique hue—pink or beige or some unconventional color. But no. Though the mind be converted, the spirit revolts at such heresy.

So in the case of a radio, a kitchen mixer, a hundred new products that would seem to cry out for a singular color treatment—the old favorites remain inviolable. Many a manufacturer who had a revolutionary product has seen it gather dust on a dealer's shelf because it wasn't the right red or blue—despite all protestations as to its remarkable and astounding utility. Experience has shown again and again that form and function are not to be seen apart from color. Appearances, superficial or otherwise, are at all times vital.

The matter of design doesn't seem to stir much recalcitrance in human beings. Make it eighteenth century, Federal, Swedish primitive, modern, and surround it with a bit of romanticism, and the average head will lend nodding attention. There will be one reservation, however—let not the design be too abstract or too unreal. A rose by any other shape will never sell so well. But make the flowers or the plaids or the geometrics reasonably sensible, and they may be submitted under almost any guise of period or vogue. Likewise, reform the aspect of a lamp, a washing machine, or a plow; increase their utility, and the new shape may prove a real incentive to purchase.

Not so in the case of color. Here the liberal bearing of the consumer may change to haughty defiance. You are getting too personal. It is your privilege to take liberties with patterns and proportions and lines; but color is her endowment, and you would do well to respect it.

These observations are well confirmed by sales experience. To refer to a survey among dealers in linoleum and colorfully printed hard-surface floor coverings, the large majority of consumers enter a store having a definite color in mind. A woman is not inclined to be fussy about design. Let her prefer a particular pattern but not its hue, and she will refuse to buy. But let the color be pleasing, and any objections she might have to the design will promptly melt away. You may switch from one design to another if the color is right; but no amount of pleading or coercion will effect a shift from one color to another, regardless of design.

Human Heritage

There are a number of accountings for public taste in color. Several of them will be discussed in this book. Some trace from history, others from psychological research, still others from the very physiology of the human eye, brain, and body. Yet the dust raised by those who whack about at spiritual and aesthetic values very often clouds an easy perception of fact. The artist will say that color and feeling are one, that some weird gyrations of the spirit are to be given credit. The businessman, rather lost in the dominions of beauty, will sit at one side and let his more particular associates decide for him. So the stylist effuses, while the businessman perspires. In the constant struggle to do things differently and to perform tricks that outsmart competition, one mistake after another is made and unsold inventories pile up in a heap. It is as though industry would fashion a dress for its public by standing off at a distance and guessing about measurements. What this person or that person thinks is to be the guiding rule. That the body of society is entirely capable of being weighed and of having its dimensions encompassed by a tape measure seldom enters into the reckoning.

Yet color is less an enigma than is commonly supposed. When it is studied not as individual feeling and expression but as mass psychology and mass reaction, its mysteries are rapidly unfolded and its harsh realities exposed. You may become a great artist by offering your own creative impulses at the shrine of humanity; but you will hardly sell merchandise on the same principle. Far more than prophecy, industry needs aesthetic socialism, an ardent attempt to glorify and enhance those urges for color which are inherent in all beings. People are inarticulate; their tastes are simple. They walk in an endless quest of beauty. You touch their emotions—and their capital—when you give expression to their fancies, when you serve their desires and not merely your own.

And this, be sure, constitutes great art, tested by any worthy standard.

When the history of color is reviewed, many valuable lessons are learned. Every so often, a generation makes up its mind that richness and purity of color are a vulgarism and that proper taste will recognize no such simplicity: good taste is subtlety, delicacy, refinement. For high-fashion markets it most assuredly may be; but the tragedy of this belief is that the loud ecstasies of the few too often lead the manufacturer to heed the cry and to suffer accordingly for it in mass markets. Track down almost any merchandising failure and it will prove to be a pink elephant that has wandered, not from the cottage, but from the mansions of good taste.

The lesson never seems to be learned. Culture is not what sells color in big volume. The longing within people is not so pedantically fashioned. Yet, despite very reliable evidence, the inspired guesswork goes on and the creations of stylists schooled to fancy notions continue to be rejected by humble people who will have none of them. In an effort to lead the parade there is a lot of rushing down blind alleys only to find that the main column of marchers has gone off in another direction.

When one studies the history of color, much is learned of people. This scholarship, however, is seldom a habit of the artist or stylist who prefers to rest on his own conceit. But the effort is wise training and will serve to reveal many of the facts and elements that make up human psychology.

Of immediate enlightenment is the discovery that mortal love for color did not arise from some primitive hunger for an aesthetic opiate. If anything has been practical to man it has been the art of color. There is a paucity of evidence that man, once he came out of a cave to build a house, felt the urge to make his environment pretty and ornate. (This attitude, in truth, didn't come into fashion until well into the fifteenth century A.D.) What attracted him to color was the mystery of his being, his love of life, and his fear of death. God fashioned the hues of the rainbow; what meaning might they have?

Thus, one finds the early beginning of a complex symbolism, which related color more to the Mysteries than to artistic fancy. The Egyptian, who considered himself an eminent member of the red race, applied dye to his flesh to emphasize the distinction, thereby founding the business of cosmetics. In like manner, he used gems and precious metals to fashion amulets having definite and specific merits in curing his ills, assuring abundant crops, safeguarding him in battle, and transporting his soul safely across into the infinite—thereby founding the craft of the jeweler.

To the ancient, colors marked the four quarters of the earth, a symbolism recognized in China, Greece, Ireland, and America. In America these color associations existed in the mythology of practically every Indian tribe. Thousands of years ago, according to one fable, the Navahos dwelt in a-land surrounded by high mountains. The rise and fall of these mountains created day and night. The eastern mountains were white and caused the day. The western mountains were yellow and brought twilight. The northern mountains were black and covered the earth in darkness, while the blue mountains to the south created dawn.

The American Indian also had color designations for a lower world, which was generally black, and an upper world, which had many colors. All this symbolism was a part of his art. The tattooing on his face, as well as the colors on his masks, effigies, and huts, were full of meaning and not merely the products of an artistic temperament. He applied these hues of the compass to his songs, ceremonies, prayers, and games. Even today the Hopi in executing a dry painting is a mystic before he is an artist. Religiously he places first his yellow color, which represents the north. Then, in order, he places the green or the blue of the west, the red of the south, and lastly the white of the east.

In European and Asiatic civilizations the conception of the universe as comprised of simple elements was also associated with color. The Hindu Upanishads, dating back to the seventh and eighth centuries B.C., recognized red as the color of fire, white as the color of water, and black as the color of earth. To the Greeks blue was for earth, green for water, red for fire, and yellow for air. The Chinese, recognizing five elements, related yellow to earth, black to water, red to fire, green to wood, and white to metal.

The planets were identified with colors and so, too, were the signs of the zodiac. Writing of the Temple of Nebuchadnezzar, which existed and was described by Herodotus in the fifth century B.C., and which has been uncovered in modern time, James Fergusson says: "This temple, as we know from the decipherment of the cylinders which were found on its angles, was dedicated to the seven planets or heavenly spheres, and we find it consequently adorned with the colors of each. The lower, which was also richly panelled, was black, the color of Saturn; the next, orange, the color of Jupiter; the third, red, emblematic of Mars; the fourth yellow, belonging to the Sun; the fifth and sixth, green and blue respectively, as dedicated to Venus and Mercury, and the upper probably white, that being the color belonging to the Moon, whose place in the Chaldean system would be uppermost."

Compare this ancient practice in architecture, in which color held special meaning and purpose, with present-day ideas of decoration, and one easily appreciates that the ancient had something else in his soul besides a lovely color effect.

The religions of the world are thus rich with a telling color symbolism. There were specific hues to identify the chief deities of the world. Yellow and gold were for Brahma, Buddha, and Confucius; green for Allah; blue for the Lord, "as the appearance of a sapphire stone."

In India the original four castes had color tokens: white for the Brahman priests; red for the soldiers and warriors; yellow for the merchants and farmers; black for the serfs and slaves. In magic, sorcery, and divination, an elaborate symbolism was developed by the ancient. Gems, bits of cloth, amulets, charms—all, with their particular efficacies—were as much a part of household equipment as the present-day bottle of iodine in the medicine cabinet.

In the practice of healing, color has many odd traditions to ponder over and many curious superstitions to think about. Because disease came mysteriously out of nature (microbes unknown), the magic of color was called upon: black threads to cure earache; scarlet cloth to stop bleeding; red flannel to cure sprains, sore throats, and fevers; yellow turnips for jaundice; amber to help the kidneys, liver, and intestines; amethyst for gout; emerald for diseases of the eyes; garnet for skin eruptions; jade for the pains of childbirth.

Strangely, many of these beliefs have survived. The efficacy of red flannel is still insisted upon by many grandmothers in the treatment of sore throat. Farmers of Europe still wear gold rings when planting their grain. Goats and cows are still to be found with red ribbons tied to their horns. Red or blue talismans still hang from the doors of houses, to stop evil at the threshold. The bride still wears "something blue," and the widow is still draped in black.

The important point to remember is that color has always been close to life. Modern business and styling err when they assume that aesthetic urges are what inspire worship of the rainbow. There is no doubt but that color is pretty. Yet the frequent failure of certain colors to sell certain merchandise has deeper significance than one might suppose. As later chapters will strive to clarify, the inherent taste of people for color is frank and unassuming. Whether the reason is to be found in the physiology of seeing, in the psychic make-up of the individual, or in the past history of color expression, makes small difference. The fact is that public taste has within it many universal qualities, responses that seem to be inherent in human nature. To put too much stress on an artistic viewpoint is to train to become a cook by dealing with herbs and spices alone.

Those who style with color today are inclined to think too much in terms of insight, feeling, creativeness. Only since the fifteenth century has man divorced color from symbolism and applied its glories for their own intrinsic sake. Take architecture and sculpture as examples. Regardless of the form of Greek architecture, the colors applied to it were generally the same, and also were generally applied in the same way to the same places. Sculpture was a matter of carving effigies that subsequently must be painted and adorned in prescribed hues. The Greek citizen was shown no vague abstractions. The decorations on his temples were pictorial history. His statues were gods or the personifications of immortals. No work of art then expressed the spirit of transportation, time, humanity, and the like, as today's art sometimes does. The Greek and his ancient neighbors did not think thus.

All this explains why practically all early color expression was simple and was comprised of simple colors. Each hue used was in some way identified with the Mysteries. The individual artist did not struggle to express his own vanity. On the contrary, it was his charge to adhere to a symbolism that all men understood. Surely, if the Greek was so sensitive to form, one might have expected him to be equally sensitive to color. However, there is little subtlety in ancient color (always a bold use of red, green, blue, gold, white, black) simply because color was a definition and not an inspiration to the artist.

Styling for the Masses

To repeat, human taste still draws from these same origins. Color styling for the masses is not a problem to be solved by high aesthetic spirit so much as by resourcefulness in understanding the average mortal. Those who have had experience in the merchandising of color well know that trends are difficult to force. You can lead people to almost any design, shape, or form; but, unless color gains a ready and almost spontaneous acceptance, it is almost certain of failure. How necessary it is, then, to pay more attention to your public and less to yourself!

It is the endeavor of this book to list those principles which will assure success in reaching mass markets through color. I say that the problem is not too difficult. Yet, before anything else, it requires that the usual thinking processes be abandoned for a more discerning viewpoint.

1. The masses of people have simple taste in color.

2. Their ideas of refinement and elegance more concern modifications and glorifications of elemental colors than they do strange or intermediate hues.

3. This means that variations of primary hues, like red, green, blue, hold more instinctive appeal than do purples, yellow-greens, blue-greens, and the like, which show too much departure from primitive qualities.

4. Originality with color is generally futile unless public acceptance is indicated. Creative styling, therefore, is better achieved through objective analysis than through personal invention.

5. Color trends in mass markets change slowly (in contrast with high-fashion markets, where shifts are more rapid). Such changes are likely to concern lighter or darker values of elemental colors, rather than variations within the spectrum. For example, a preference for red today will more likely be followed by a preference for rose or maroon or pink than by a preference for red-violet or red-orange.

6. There is never any assurance that an "exclusive" color, developed in terms of high fashion, will hold popular appeal. Mass markets do not always follow high-fashion dictates, but may move along with real independence. Fifth Avenue may often get no farther west than the Hudson River.

7. As a matter of merchandising policy, and because of the peculiarity of human temperament, it is better to offer many styles or patterns in few color ranges than it is to offer few patterns in many color ranges.

8. Because of the frankness and impulsiveness shown by most buyers in low-income groups, it is wiser to be obvious than subtle in color selection and arrangement. In other words, it is safer to have the hue a little too bright than a little too dim. (In high-fashion markets the reverse policy is most effective.)

9. In mass markets overstyling is far more dangerous than understyling. If creativeness is to be shown, let such effort be expended toward the enhancement of colors having good sales records. A "better red" will usually accomplish more than an extra hue of some sort promoted for some presumed novelty.

10. Be cautious about any aesthetic opinions. Your public has good taste, or colonial furniture and peasant textiles would never have come into existence and have survived. Be philosophical to the extent that high-fashion values in color may be more ephemeral than low-fashion values. You perform the most service, sell the most goods, create the greatest "art," when you satisfy the majority of people. This need not be a process of lowering anyone's dignity. On the contrary, even the simplest and plainest color (and design) may be perfected to a beauty that will make it worthy of any museum. The trick is to have the art expression democratic rather than aristocratic, to conceive it as a public, rather than a private, ideal.

Color and Language

There are some who hold the opinion that ordinary human beings crave individuality. They support the theory that individual personality, individual beauty, individual this and that are the crying need of the average soul. This delusion is, of course, a carry-over from the high-fashion type of thinking, where to be different is in itself a sufficient credo. Yet there is a massive weight of evidence to show that, for a given amount of effort, one right color or shape will sell more units than a dozen colors or shapes can do. From the standpoint of business economy (and how seldom this happens!), there is more profit in concentrating on few things than on many. Yet the very setup of most styling departments leads to opposite endeavors. A stylist is not inclined to make a best seller better. He studies his so-called "dogs," bothers his head over the low ends of his lines, and too frequently disposes of one specter only to summon up a few more.

In principle, sound creativeness with color attempts to increase healthy things and not alone to nourish anemic ones. To determine *not* to do certain things is often the most profitable and progressive move industry can make.

The problems of color seem to stimulate too much overthinking and not enough common sense and rationalization. It may be that I stress the point too taut, but for some inexplicable reason the simple taste of most people seems to escape comprehension. Many stylists, left to themselves, would unnecessarily complicate the plainest of facts.

Yet that people hold a modest regard for color is revealed in their own verbal attempts to describe it. If language is an expression of human hearts and minds, then, any notions about the subtlety of public taste in color is quite upset.

When people are interested in something, they give it a name—radio, Cellophane, and so on. When interest is negligible, the name remains unformed. The average person, when asked to write down all the colors he can recall, will list about 30 or 40. And out of these 30 or 40 almost half will be duplicates, as far as generic terms are concerned—crimson, ruby, vermilion, for example, being descriptive of red.

Being so inarticulate in speaking of color, people are naturally exposed in their emotional preferences. Were they at all exacting about subtle hues and modifications of hues, the dictionary would be packed full of appropriate definitions. (Even as it is, most colors have borrowed identifications: rose, violet, orange, peach, emerald, amber, etc.)

Of real interest is the fact that only about 18 colors can be named by the average person with any assurance that others will know what he is talking about! And these colors (certainly not by accident) will be found to comprise the best sellers in almost every line of mass market merchandise in existence. They are the hues that will arouse the most attention value in an advertisement or a display and sell the most packages. Here they are: red, orange, yellow, green, blue, violet (or purple), pink, buff (ivory or cream), flesh (or peach), lavender (or orchid), brown, maroon, tan, white, gray, black, gold, silver. While names such as beige, ecru, mauve, and the like may be understood by a few elect, they are largely meaningless to the public (and generally do not sell well in mass markets).

It must be that simple colors, being memorable, are therefore marketable. Odd colors, being forgettable, are the risks and speculations of business.

Having this phenomenon as a clue, the answer to consumer styling is to be tersely stated: People at large have definite color preferences. Though these preferences may change, they are almost always of an elementary order. The task of styling is to glorify these desires with every possible skill and originality. The art and business of color are on a sound and enlightened basis when the approach is objective and when human wants are carefully analyzed, measured, and anticipated.

Chapter 2

THESE ARE THE COLORS THAT PEOPLE PREFER

COLOR may be discussed from several viewpoints. When one thinks of it in terms of people, however, of human experience and sensation, there is little need to discuss matters of physics and chemistry, because they will only serve to confuse the issue. Color is greatly like flavor and odor; they all have physical origin in some sort of stimulation. Yet to analyze the chemical composition of sugar or the oil of a rose is by no means to define sweetness. The latter is a human interpretation, a psychological thing. It is sensation—and sensation rather than wave lengths or molecules or atoms is what should concern those who use color (or taste or odor) in merchandising and advertising.

Aristotle had this to say: "Simple colors are the proper colors of the elements, *i.e.*, of fire, air, water, and earth." They owed their origin to "different strengths of sunlight and firelight," a belief that was championed for some eighteen centuries. It was Newton, however, who in 1666 felt justified in concluding that white light is not simple, but is comprised of many rays, which the prism can separate.

A purely psychological attitude—color as personal sensation rather than as physical energy—was slow in arriving. In truth, only in the past century has color been intelligently analyzed and interpreted from the human standpoint, physics and chemistry aside. Science, being quite objective in its viewpoint, has not always acknowledged the fact that, despite the composition of bone and tissue and the atomic structure of all matter, there is a vast difference between a dead man and a live one. And color, of all phenomena, has sorely needed the live approach.

It is not an easy thing to define color. J. H. Parsons, an English psychologist, writes: "The untutored regard the green of a leaf as an attribute of the leaf. The physicist, however, knows that color depends upon the light reflected from the leaf and calls the reflected light green. The physiologist knows that the leaf, which appears green when looked at directly may appear yellow or gray when its image falls on the periphery of the retina. He is, therefore, inclined to regard color as an attribute of the eye itself. Finally, to the psychologist the green is not an attribute of the leaf, or of the light, or of the eye, but a psychical phenomenon, a definite qualitative entity in consciousness." In other words, color is visual experience. It is a feeling inside you, a sensation that you associate with vision and that marks the appearance of all things. Everything you see, therefore, is *color,* whether it be red or blue, black or white.

The Elements of Color

In the middle of the nineteenth century, James Clerk Maxwell, a celebrated physicist, wrote, "The science of color must . . . be regarded as a mental science." This attitude, adopted by many scientists, marked the beginning of pure psychological investigation, which was to divorce color from physics and wed it to human consciousness, a more soulful mate. Since then the science of psychology has grown and its investigators have brought forth a mass of evidence that offers practical aid to industry in its color problems. Unfortunately, much of this research has been disregarded by the artist, who so habitually turns a deaf ear and a blind eye to things scientific. The psychological aspects of color, however, while technical, are nonetheless based on the very humble study of human consciousness. And as color and people have so much in common, there is every benefit in perusing the course of psychological inquiry and then later checking these findings against the hard facts of market and sales data.

In the spectrum are innumerable wave lengths of light. As far as radiant energy is concerned, each one of these is as significant as the next. Yet when the human eye looks at the spectrum (or the rainbow), these wave lengths are automatically sorted out into neat bunches, each having a sort of family resemblance. Thus, Selig Hecht writes, "Color vision may be defined as the capacity of the eye to divide the visible spectrum into a series of regions which produce qualitatively different sensory effects."

Why the human eye and brain, seeing *many* wave lengths, should sort them out and interpret them as representing *few* sensations is something of a mystery. However, one of the first of the new truths is thereby revealed: it is human to see color in simple terms, to be attracted to major regions rather than minor parts in the spectrum. This fact has direct application to merchandising and will be emphasized in later chapters.

To the psychologist there are six primitive or primary sensations: red and green, yellow and blue, white and black. These colors are unique, and they bear no resemblance to each other. This, again, is somewhat mysterious but perfectly evident in human experience. Variations in the sequence, red, yellow, green, blue, and back to red are hue variations. An intermediate color like orange will look something like red and something like yellow. But red and yellow will not necessarily look like orange. So with violet, which may be comprised of red and blue.

A further observation is that all modifications of a single hue may be represented by a triangle, with pure color on one angle, white on the second and black on the third, for all perceptible colors may be formed by graded mixtures of these three elements.

This is color sensation! It has nothing to do with the physics of light or the chemistry of pigments. It is human psychology, the way in which the eye and mind react to visual stimuli.

Look at the illustration of the Triangle. Consider this a diagram and a summary of how you and all other persons see color.

1. First of all, you distinguish the quality of hue. The unique parts are red, yellow, green, and blue, which, in mixture, produce all the *pure* colors of the spectrum. Pure colors represent one primary form of sensation.

2. The second primary form is white. White in sensation is also unique and does not in the least resemble any pure color (red, yellow, green, or blue).

3. The third primary form is black. And, again, black looks nothing like white or any pure color.

4. When these three primary forms are combined, four secondary forms are produced, as indicated on the Triangle.

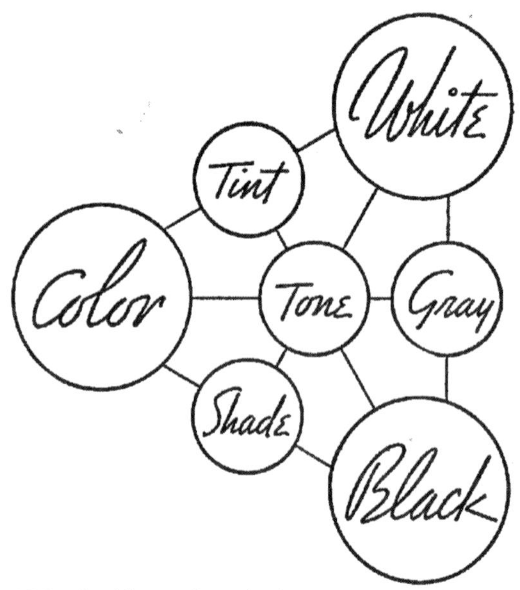

The color Triangle. All sensations of color are to be classified in seven forms.

White and pure colors produce what are called tints—whitish colors, such as pink, lavender, peach. Such tints appear to have both pure color and white in their make-up. When black and pure colors are combined, shades are produced, such as brown, olive, maroon. These, as well, appear to have both black and pure color in their composition. Black and white similarly produce gray, another unique and intermediate type of sensation. Finally, a combination of all three primaries—pure color, white, black—produces what is called a tone, such as tan. The tone, incidentally, becomes the most neutral of all color forms.

These points may seem academic. Yet, realize that a clear conception of them has taken a considerable time. The scientist has had much difficulty in divorcing color as sensation from color as energy—wave lengths and atoms.

To be practical, one must understand that color from the human viewpoint settles down to an all-embracing simplification; for, despite all subtleties and variations seen by the eye, every color sensation can be classified as one of these seven forms! These are the seven boxes of the mind, and into them go a multitude of visual experiences to be given real sense and order.

It naturally follows (to be confirmed by the merchandising of colors in mass markets) that the colors people like best are those which stand as the noblest representatives of the seven forms. This means vivid reds and blues, sheer whites, deep and concentrated blacks, clean-cut pastels, rich shades, soft grays. While these may appear to include almost everything, they are not so many in number as might be thought. Every lucid example of form in color may be matched by dozens of examples of borderline colors, off shades of white, black, and gray, odd tones of intermediate hues—high-fashion colors all, appreciated by cultured taste, but refinements that never seem to appeal to the masses (for apparently good scientific reason).*

Color Preferences

At least 50 authoritative tests have been made of human color preferences. The literature is so complete and the results are so uniform that one is hardly able to question the conclusions reached. Should the reader be interested in the

* It is always convenient to think of color in terms of pure color, white, black, gray; tint, shade, tone. Each of these is fairly specific in its definition and will help to clarify any discussion of the subject of color. Another series of terms used by the scientist and the technologist consists of hue, lightness, and saturation; or hue, value, and chroma (Munsell). Hue is the quality that distinguishes spectral character, red, orange, and so on. Lightness (sometimes brightness) and value refer to the relative lightness or darkness of a color as compared with the steps of a white-gray-black scale. Saturation, purity, and chroma are terms that refer to relative degree of departure from a neutral gray. Thus, pink is a red color of high lightness or value, maroon is a red color of low value. Bright red has full chroma (saturation or purity), while a soft, grayish rose (possibly of the same hue and value as pure red) would have weak chroma.

details of such investigations, let me refer him to two excellent articles, one by J. P. Guilford in the *Journal of Experimental Psychology* for June, 1934, and the other by H. J. Eysenck in the *American Journal of Psychology* for July, 1941. Here he will find comprehensive reviews, and references by the score.

To give some order to this matter of color predilection, consider first the reaction of babes. In the first months of life it is difficult enough to learn to see, to fix both eyes on the same object, and to make sense out of visual experience. Some authorities declare that touch and form are dominant in infants, and that a real love for color does not become evident until well after the second year. Then color perception begins to rival form perception. R. Staples exposed disks to infants and measured the duration of visual concentration. The babes looked longer at colors than they did at neutral tones. Their favorites, judged by certain eye fixations and reaching efforts, were red first, then yellow, green, blue. Apparently the infant is most attracted to brightness and richness of hue.

In small children, a liking for yellow begins to drop away—and to keep dropping with the years. Now the preference is for red and blue, the two universal favorites, which maintain their fascination throughout life. The order in childhood, therefore, is red, blue, green, violet, orange, yellow.

With maturity comes a greater liking for hues of shorter wave length (blue, green) than for hues of longer wave length (red, orange, yellow). The order now becomes blue, red, green, violet, orange, yellow. And it remains thus, the eternal and international ranking.

That color preferences are almost identical in human beings of both sexes and in persons of all nationalities and creeds is substantiated on every side.

T. R. Garth found that American Indians preferred red, then blue, violet, green, orange, yellow.

Among Filipinos, the order was red, green, blue, violet, orange, yellow.

Among Negroes the order was blue, red, green, violet, orange, yellow—the same as for practically everybody else.

Even among insane subjects, S. E. Katz found almost the same rankings—blue, green, red, violet, yellow, orange.

Green was best liked by male inmates, and red by female. Warm hues seemed to appeal to morbid patients, and cool hues to the more hysterical ones.

There seems to be good reason, therefore, for one authority to say that there is "sufficient agreement upon color preferences to indicate a basic, biological cause of likes and dislikes for colors."

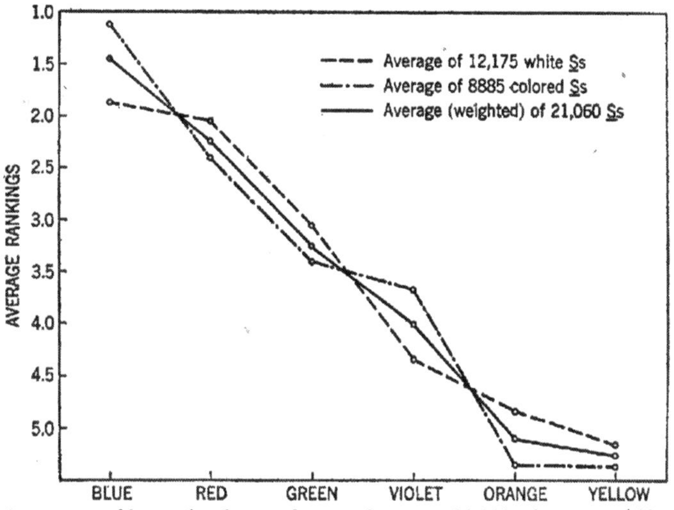

Average rankings of color preference for over 21,000 observers. (*After Eysenck.*)

(It so happens, however, that, while red and blue are always predominant, blond racial types generally prefer blue and brunet racial types generally prefer red—a point that will later be discussed, when an attempt will be made to explain why.)

To summarize the whole picture, Eysenck tabulated all research involving some 21,060 individual tests. Blue ranked first, then red, green, violet, orange, yellow. In a similar recapitulation for sex differences, the order was the same, except that men put orange in fifth place, yellow in sixth, while women put yellow in fifth place with orange in sixth. (To the ladies, yellow is not quite so bad as orange.)

It must be remembered, of course, that the color-preference tests of the psychologist are for so-called absolute choices—color for the sake of color, and without reference to palpable things. A later chapter on sales records will show what happens when the selection of hue has to do with products and when practical considerations, rather than mere aesthetic fancy, come into play.

This scientific work, however, is truly valuable, not alone for the hue rankings that it divulges and for the universality that it finds in human love of color, but for the many sidelights that it throws on emotional propensities. Here, for example, is a further list of research facts that have direct bearing on the problems of selling color and influencing people.

1. In absolute choice, pure, rich colors are generally preferred to modified ones.

2. The three most preferred hues are blue, red, and green; the three least preferred are violet, orange, and yellow. These are the so-called maxima and minima—and all intermediate, or "mixed," pure hues are not liked so much.

3. However, one blue (or red or green, etc.) is not necessarily equivalent to another in "pleasantness." This means that every region in the spectrum has its best example, which will, to be commercial, sell better than its immediate adjacents.

4. Although different backgrounds will affect the visual aspect of colors, this does not seem to have any serious influence on color-preference rankings.

5. The major interest of people is pure color. This quality is always more dominant than whiteness, blackness, or grayness in a color. In fact, modified tints and shades follow the same general order of preference as for pure colors.

6. Saturated, or pure, colors, however, appear to be best liked in small areas. In larger areas a preference is indicated for lighter tints.

7. Colors tend to appear stronger as their areas are increased. Judgments of value or lightness, however, remain fairly constant, regardless of area.

8. Women are more inclined to be tolerant of modified colors than are men, who tend to be more loyal to vividness. Guilford writes, "It would appear that both men and women

prefer the lighter colors to the darker, and that this tendency is stronger for women than for men, and that both prefer the more saturated to the less saturated, the effect of chroma being more strongly felt by the men."

9. There is some evidence of two distinct systems of color preference, one being more primitive than the other—certain people preferring saturated colors and others preferring softer tints and shades.

10. "On the average, agreement between rankings of colors is as high as agreement between tests of intelligence" (Eysenck).

Color Combinations

A good deal of research has also been devoted to the beauty of color combinations. Here, however, the psychologist has less to contribute to the problems of styling. It is the curious notion of many technical-minded investigators (a notion based upon their controlled tests) that the beauty of one color may be enhanced through juxtaposition with its complement. As a matter of fact, its appeal may merely be compromised. On this point Guilford has again offered the best and most practical guidance, and his conclusions will be given on following pages. Good styling is nearly always a process of glorifying one thing. While so-called balanced color effects are often suitable for packages, displays, and posters—because of their startling qualities in vision—they too often fail at selling consumer merchandise, probably because the buyer's dominant preference is confused and a choice of heart is distracted.

In working with children, M. Imada found that color preference was not haphazard, even though good discrimination was not as yet highly developed. Given black crayons, the youngsters were inclined to draw concrete things, vehicles, buildings. When the same children were given colored crayons, their fancies were more inspired to attempt human beings, animals, and plants. Red with yellow and red with blue were favored combinations.

In similar experiments, Ann Van Nice Gale found yellow popular in combination with red-violet or blue. The combination of blue and green also was liked. Contrast, naturally, was more exciting than analogy or subtlety.

In testing adults, using colored lights thrown upon a screen, William E. Walton and Beulah M. Morrison found the combination of red and blue highest in ranking, then blue and green, red and green, clear and blue, amber and blue, amber and green, red and amber, with clear and amber last.

Murray and Spencer, in experimenting with the appeal of color on different backgrounds, found black to be the best ground, green next, and red the least desirable. (A large area of red would naturally put most other colors to shame.)

The Work of Guilford

J. P. Guilford, whose work in the field of color preference has been mentioned, has in the author's opinion conducted the ablest and most helpful of all investigations into the affective value of color. Too frequently the scientist has worked with a limited number of hues (a dozen or so), and these usually in full saturation. It has been Guilford's interest to see what might happen when the selection was far more diverse. He also wonders if the preference indicated for pure colors holds true when qualities of lightness and saturation are introduced, when the choice is among colors that are not so brilliant as the proverbial rainbow?

In styling, it is not only vital to know what major colors are best, but also to find out what particular variations are most pleasing. Does the tinting or shading of a pure color result in increased or decreased appeal? Although red, for example, is universally liked, what about pink and maroon? And what particular pinks and maroons are superior? Answers to questions such as these are naturally important in merchandising.

In a report published in the issue of *Psychometrika* for March, 1939, Guilford describes his efforts. He selected 316 colored papers that matched systematic samples of pages from the Munsell *Book of Color.* Here are some of his findings.

With a variety of colors, all *very* grayish in saturation and shown in steps from a medium deep to a medium light tone, (*a*) the lighter tones were all liked better on an average, (*b*) but only the lighter blues and violets were rated as truly pleasant.

When similar scales of colors having a generally richer cast (rather than an extremely weak cast) were shown, far more colors were rated as pleasant, particularly in the region of green.

With the scales further increased in saturation, pleasure in blue, green, and violet seemed to pick up; while yellow, orange, and yellow-green remained lower in appeal.

At very high saturation, the preferences naturally agree with those observed by practically all investigators who have worked with pure colors.

The above tests thus serve to compare the relative beauty of an assortment of pure colors against grayish ones and to reveal the natural liking for richness over weakness in color.

A second approach of Guilford was to determine what happened when colors *of the same hue* were exhibited in various tints, shades, and tones. Here are further results.

With grayish tones as against pure tones, the purer forms are liked better.

With light tones as against dark tones, the lighter tones are liked better.

These observations seem to apply to all individual hues, with the exception of yellow. Dark, grayish yellows seem to have more appeal than deep, rich yellows (which are, in reality, mustardy and olive in character). Light, grayish yellows, however, will be preferred as they approach clarity. Very pale yellows are liked best of all.

Guilford also observed that cool colors are likely to find their best appeal at low levels, yellow and orange at high levels. Reds are also preferred as they become lighter.

With a color such as blue, a pure tone is well liked at a fairly dark and saturated level. In a grayish form of blue, however, a light tone is preferred to a deep one.

"We might suggest as a general rule that . . . there is a tendency for colors to be preferred at tint levels where they can be most saturated." Thus colors normally light (orange, yellow) are pleasing in pale tints (ivory, buff, peach). Colors normally dark (blue, violet) are pleasing in deep tones (navy, wine). Variations of red are more likely to be erratic.

These points may be a trifle difficult to grasp, for Guilford's control methods have rather strict order insofar as scientific color organization is concerned. He goes about an important business. To the author's knowledge, he is the first man ever to make a practical and comprehensive study of color harmony in all its variations of tint, shade, and tone—a study that industry is forced to cope with every hour of the day. An analysis of sales records, for example, is comparable to the thing Guilford is doing. In Guilford's case, however, the larger world of color experience (all neatly plotted and arranged) is surveyed, as against the more or less heterogeneous collection of samples considered in most styling efforts.

Guilford has drawn charts to explain these "affective values" of individual hues, and one of them is illustrated herewith.

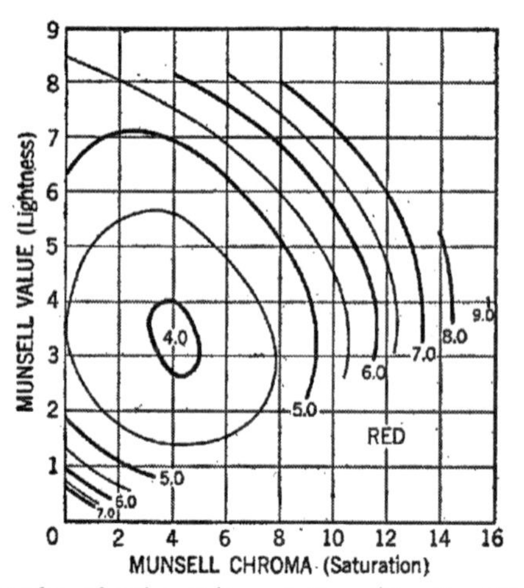

The affective value of red and its variations. The line 5.5 indicates the division between likes and dislikes. Colors to the right and left are pleasant and unpleasant. (*After Guilford.*)

It is very likely that a complete set of these charts will be published in the future. They would prove useful indeed and would offer industry a very neat and very comprehensive picture of human likes and dislikes.

The chart reproduced is for the color red, and the arrangement is according to Munsell. Tints, shades, and tones, as well as the pure hue itself, are included. To quote Guilford, "For any color sample whose hue is red, having determined its Munsell value and chroma, one can easily read off its most probable average value for liking or disliking. Thus, a red of Munsell value 7 and chroma 10 (briefly designated as R 7/10) would have an expected affective value of 6.8, and Munsell R 3/12 would have an expected affective value of 6.3. The heavy line at 5.5 indicates the division between likes and dislikes. Color samples to the right of it are all pleasant; those to the left, with the exception of the colors very close to black, are unpleasant."

Thus for this particular hue, bright pinks and strong, pure reds are most appealing. The weakest and most unpleasant region (around 4.0) consists of grayish reds of medium or medium-deep value.

As to his charts, Guilford writes: "The practical value of these isohedon charts should be very apparent. Assuming that we can obtain in this manner the intrinsic affective values of colors for the masses of buying customers, it should be relatively simple to set up a series of charts, one for each of the twenty Munsell hues, let us say. Once any particular color sample is evaluated on the Munsell system, a glance at the appropriate chart would tell how well the average person likes it. Predictions for single individuals cannot be so accurately made as for groups, of course, but in these days commodities are made to please the masses."

Chapter 3

SCIENCE OFFERS AN ANSWER

GUILFORD writes, "I think that it is more than a figure of speech to say that living tissue, particularly brain tissue, generates colors and pleasantness or unpleasantness just as other collections of matter generate the phenomena of heat, or magnetism, or electricity." How else can one account for color preferences?

While science pretty well disregarded the psychological and human aspects of color for many generations, it recently has begun to acknowledge that the beauty of color may arise from more corporeal sources than spiritual or aesthetic feeling. Assuming that there may be "something in the glands," many investigators have set to work, and the results have been exciting indeed. The difference between the live man and the dead one is finally being determined.

There are good causes and reasons for color likes and dislikes, for racial preferences, for dozens of reactions that make color the most fascinating of all natural phenomena. This material has been hard to get. It is found in widely separated places, both here and abroad. It has seldom been assembled, reviewed, and interpreted. You cannot find a complete book about it (although the author now has one in preparation). Yet, the attempt to explain color in biological rather than esoteric terms, to trace its reality in sensation to things physical and physiological, is proving to be a valuable contribution to the cause of color, commercial or otherwise; for it gets inside the clock, so to speak, and sees what makes the chimes ring.

Understand that the *explanation* of color preference is new to science, so new in fact that you will not find one color authority in a hundred who can speak with any coherence about it. Ask yourself *why* you like blue (if it so happens that you do). Then realize what a difficult but intriguing problem it would be to write a sensible or even plausible answer.

Light and Life

Because light is essential to all living things, one may naturally expect color to hold real significance. In the lowest forms of animal life the humble amoeba makes its way in and out of light, "seeing" with its entire organism and generally choosing a shady spot, neither too bright nor too dim. A more advanced cousin, the euglena, with light-sensitive cells clustered at the base of its feelers, wriggles here and there for that degree of brightness best suited, no doubt, to its existence.

The starfish, with eyespots on its arms, can "see" through its skin if these "eyes" are somehow nipped off in battle. The earthworm, nocturnal in its habits, answers the warnings of receptor cells scattered on the walls of its body and stays clear of daylight, unless it is flooded out of the ground. The clam, with "eyes" on the inner lip of its siphon, retreats quickly into its shell when a shadow passes. And so it goes, on up the scale of light sensitivity.

Sensitivity to color, however, is to be found on higher rungs of the ladder of life—in the turtle, the octopus, the fish, the snake—reaching high perfection in the bird; but for some reason or other, it is missing again in most mammals, such as the horse, the cat, the dog (but not the ape).

Among insects, Bertholf and other authorities have found a range of vision quite different from that of man. Whereas man sees that span of electromagnetic energy extending from red to violet, the insect begins to see at yellow (red and orange being mere darkness); and it keeps on seeing through green, blue, into ultraviolet and beyond. E. N. Grisewood had fruit flies responding to wave lengths almost as fine as X-rays, radiant energy with which the insects surely had no experience in nature.

Again, one finds light regulating certain familiar habits of crawling and flying pests—the cockroach scampering away as fast as it can, the moth throwing caution to the winds.

This is light, however, not color specifically. Yet color, too, is important and has different attraction depending on its brightness and its wave length. Ants placed in a box that is illuminated by a full spectrum will carry their unborn progenies (always kept in darkness) from the ultraviolet and violet region into the red and infrared. Not "seeing" red, the ant will perhaps be under the delusion that his retreat is to a dark and secluded place.

Experimenting with night-flying insects, L. C. Porter and G. F. Prideaux found that brightness was a dominant factor in attraction power. Next to this, the more a source of illumination approaches the blue end of the spectrum, the more insects it will gather; and the more it approaches the red end, the fewer it will gather. "The substitution of yellow lamps for white lamps of equal candle power reduces the number of insects attracted by approximately 50 per cent." Blue, consequently, is the preferred hue, and red or yellow, the least noticed.

For daylight traps, however, Frederick G. Vosburg reports, "For some reason a yellow trap will catch more Japanese beetles than any other color." This probably is not a contradiction, for a yellow surface in daylight has a high relative brightness, and brightness, remember, seems to be an effective lure.

Plants

A similar discrimination seems to be shown throughout plant life. The generally green color of most plants must, of itself, represent some basic law of preference in nature—possibly, the need of the plant to absorb energy from the orange-red part of the spectrum. At least, extensive research divulges that plants thrive differently on different types of radiation.

While research in the growth of plants has been going on for well over a century, most of the early inquiries were prejudiced by an almost occult conviction of magic. In 1918, however, F. Schanz noted that plants grew taller as the blue and ultraviolet end of the spectrum was cut off. He raised plants in beds covered by sheets of glass. Cucumbers and chrysanthemums grew tallest under red light; potatoes and beets were weakest in yellow light, their fruit largest and healthiest under blue-violet light.

H. W. Popp in 1926, using colored lights of equal intensity, found that red promoted stem elongation, while blue checked it. N. Pfeiffer, under more controlled conditions, found that radiation from the visible spectrum (ultraviolet and infrared excluded) caused the greatest diameter and height of stem. Leaf development was greatest under blue, root development was greatest when the violet end of the spectrum was missing.

Dr. R. B. Withrow also found that for many plants orange-red light stimulated the most response. Under yellow, green, and blue, certain plants did not grow tall nor did they flower, although the foliage was rather abundant. (Some authorities feel convinced that blue radiation excites greatest response in plants having reserve food supplies—in tubers and bulbs.)

W. H. Hoover found red most productive in the growth of wheat—but not infrared or ultraviolet, which "contribute nothing to the assimilation of carbon dioxide in wheat." Lewis G. Flint promoted active germination in lettuce seed with red, orange, and yellow rays; A. Creslas did likewise with grass seed.

What such control means commercially is well expressed by Dr. Withrow. He states that supplementary light and color may be used to cause earlier and increased flowering in such plants as the aster, Shasta daisy, pansy. Delayed flowering may be produced in the chrysanthemum. In Europe, gorgeous flowers and big juicy strawberries have been grown commercially in hothouses where daylight is supplemented by red neon light. The advantage from the standpoint of business is either to produce a bigger and better bloom or to get into the market a few weeks before competition, and thus command a handsomer price.

Birds, Animals, Human Beings

I do not mean to go too far afield from the chief essay of this book. However, I feel sure these notes on colors, insects, plants, animals, and human beings are interesting and will help to offer more concrete evidence as to the powers of color, both physical and psychical. While a later chapter on the psychology of color will discuss some of the tangible influences of color on the human organism, let us proceed here with additional data bearing on probable reasons for normal likes and dislikes.

I suppose it would hardly be fair to say that color preferences exist among lower forms of animal life. Yet the fact that certain creatures respond differently to different spectral energy is something to think about. Who knows but that human emotional reactions may have their basis in some primitive instinct or reaction that is quite apart from so-called intelligence? Regarding the sense of vision itself, Walls, in his monumental book, *The Vertebrate Eye*, offers this statement: "Color vision itself is a potent aid to visual acuity in its broad sense, and was certainly evolved for this application rather than for the aesthetic ones which it has come to have in human vision." If color is a gift of the gods, it is a practical one concerned with survival first, and delight second.

Cora Reeves, in experimenting with mud minnows, noted that the respiratory rate did not change much in natural light, regardless of its intensity. However, when she slipped a piece of ruby glass over the minnows, they promptly settled to the bottom, had fits of trembling, and more than doubled their breathing rate. Yet one is not to jump at any quick conclusions. Color "preference" or rejection, by fishes or by other creatures, is not always the same. Walls states, "Fishes generally seem either to shun red, or to prefer it decidedly. This paradox does not appear to have interested the investigators in this field; but, granting that the red is seen as such, red-shyness and red-love both seem to indicate a high attention-value for red." Perhaps we are not greatly different from the fish—only, more articulate.

K. S. Lashley was able to train gamecocks to distinguish between red and green. In a similar study of pigeons, tests revealed that at least 20 different spectral hues could be seen. Among birds, color sense is said to be particularly well developed for the red end of the spectrum (quite opposite to the vision of insects). H. Klüver trained a squirrel monkey to distinguish the same relative color range as man. One monkey was found with an apparent emotional preference, favoring any color mixture that contained red.

In man the response to color involves many complex factors, few of which have been thoroughly analyzed. It is believed, however, that spiritual or aesthetic "feelings" may someday find a scientific basis in things physiological. Kurt Goldstein writes, "It is probably not a false statement to say that a specific color stimulation is accompanied by a specific response pattern of the entire organism"—which is the happy springboard for a plunge into the mysteries of human color preference.

To begin with, the affective values of color have three so-called maxima in blue, red, and green, and three minima in yellow, orange, and violet. The three maxima come pretty close to the primaries generally recognized in certain color theories championed by the physicist. In fact, the retina is thought by some authorities to contain nerves that respond basically to blue, red, and green. If this be true, there may be definite physiological reasons for the preference that most people show for these three hues.

To speak of real fundamentals, E. R. Jaensch came forth some years ago with the theory that the human eye originally was a brain organ associated with thought and instinct. Only later, in the course of evolution, did it devote itself to the job of recording light stimuli. In a crayfish, for example, the influence of light and color are said to affect the pituitary gland, which in turn bestirs the hormones of the creature and gets it to expand or contract its body cells—thereby producing a color change over its entire body. This means, precisely, that something of the sort probably transpires in the human organism. The eye receives the light and transmits it to the brain. The brain, in turn, telegraphs to certain glands, which then get busy and cause reactions throughout the entire vascular system.

Some men, perhaps more enthusiastic at heart, will argue that a physical response to color doesn't even need the human eye! Jules Romains some years ago (with the applause of Anatole France) demonstrated what he called paroptic perception—the ability to see form and color by sealing the eyes and exposing the neck, throat, forehead, or chest. More credible, perhaps, is the observation that a blind man will respond to light and feel its stimulation, even though non-visible heat rays are absent.

It is undoubtedly true that the body has a radiation sense, for the different wave lengths of light will vary in their penetration of human tissue. Violet and ultraviolet radiation is said to act on the superficial layers of the skin; the region from blue to red acts on the blood; while the radiation of red and infrared affects the deeper layers of flesh.

Metzger, Goldstein, and other investigators have noted a change in the tonus of the body (a condition of tenseness in the muscles) under different colors. When the arms are stretched straight out in front of the body and the eyes are blindfolded, exposure to red will cause the arms to spread away from each other. Exposure to green and blue will cause the arms to approach each other in a series of jerky motions—these actions not being consciously noted by the subject himself.

Goldstein writes, "We find that green favors performance in general, in contrast to red. The effect of red probably goes more in the direction of an impairment of performance." Interpreting this statement, Goldstein believes that green more or less lets the body be itself, lets it act more spontaneously, while red causes more of an outside attraction. Green and blue thus become the hues of the introvert, red and orange the hues of the extrovert.

S. V. Kravkov, a Russian scientist, has more to add. He says, "We may consider it an established fact that the color sensitivity of our . . . vision is dependent in a definite way on the condition of the autonomic nervous system." Kravkov, through extensive tests, discovered that the sensitivity of the eye is to be influenced by stimulation of the other senses. Loud noises, heat, sugar, all seem to increase the sensitivity of the eye to green and blue colors, and to decrease sensitivity to orange and red. The neutral point, where no change is to be noted, lies in the yellow region of the spectrum. Kravkov points out that fluctuations in the color sensitivity of the eye may be connected with seasonal changes in the functioning of the endocrine glands. If so, one may here find an explanation for the increase in preference for cool colors in summer and for warm colors in winter, observed in the sale of products to the consumer.

In similar investigations Frank Allen and Manuel Schwartz (in the *Journal of General Physiology*, September, 1940) came to this conclusion: "It may be safely inferred that stimulation of any sense organ influences all other sense organs in their excitability." Here are some of their findings.

When the retina of the eye is stimulated, sensitivity to red is depressed, and sensitivity to green and violet is enhanced. This is the immediate result following retinal stimulation. However, after a rest period of 3 minutes, the effects were reversed, red appearing to have more brightness and green and violet colors being diminished.

In the case of noise, Kravkov's findings were confirmed. "The red color of the spectrum appears of lowered intensity, the green of enhanced, and the violet of lowered intensity."

Regarding odors, the same sort of effect was noticed. "With the odor of oil of geranium as a stimulating substance, the red and violet sensations . . . were depressed in sensitivity and the green enhanced."

For a reaction to taste, sulfate of quinine was placed on the back of the tongue. Once again, "the red sensation is depressed and the green enhanced in sensitivity." A rest period of 3 minutes again produced a reversal, as it did with visual excitation (noticeable also in sound and odor stimulation). Now red is enhanced and green is depressed.

Allen and Schwartz offer several interesting conclusions. "Perhaps all sense organs are so interrelated that stimulation of any one of them influences all others." If so, many phenomena once explained in terms of nerve responses on the retina may require a different line of reasoning. For example, the fact that the contrast of one color against another will change the appearance of both may have more to do with brain processes than with those that are purely retinal.

Blonds and Brunets

Not to wander too far from the problems of merchandising color, consider now an explanation of color preference by racial type.

E. R. Jaensch, in his book *Eidetic Imagery*, mentions the difference between a predominance of "sunlight" in the more tropical regions of the world, and of "skylight" in the more polar regions. As one travels from cold to hot climates sunlight increases and skylight decreases. Intense light requires sun adaptation, or "red-sightedness," and this may be accompanied by a strong pigmentation on the foveal area of the retina.

Now, red-sighted persons are typical brunets—the Latins. They are likely to have dark eyes, hair, and complexion. Their natural preference is for red and all warm hues, a predilection which may be far from spiritual in origin and is perhaps due to a physiological process of accommodation to long waves of light.

Blonds, on the other hand, are green-sighted. They are the Nordic and Scandinavian types, with bluish eyes, light hair, and light complexion. Their preference is for blue and green.

This much is known in business—that likes and dislikes for color vary in different regions of the country, and also as populations vary. The best sellers in a Swedish neighborhood will differ from the best sellers among Italians. In the main, the conjecture of Jaensch applies, the brunets liking red and colors of strong chroma, and the blonds liking blue and soft tones in general.

There is, however, more to the story. Sunlight rather than temperature seems to be a significant regulator of human taste in color. Interesting evidence along this line has been made known to the author by Helen D. Taylor.

Refer to the map shown on page 38. This shows the relative hours of sunlight to be expected in different parts of the United States. The more cloudy districts are in the far Northwest, in western Michigan and in and around Ohio and western Pennsylvania. The sunny lands are in southern California, Arizona, New Mexico, and the southern half of Florida.

It will be noted, however, that the distribution of sunlight (or the lack of it) does not follow horizontal paths from north to south. New England has just as much sunlight as parts of Tennessee, Alabama, Mississippi, Louisiana, and Texas, and about the same as San Francisco.

While it is usually hearsay that color preferences up north differ from those down south, this is not altogether true. To

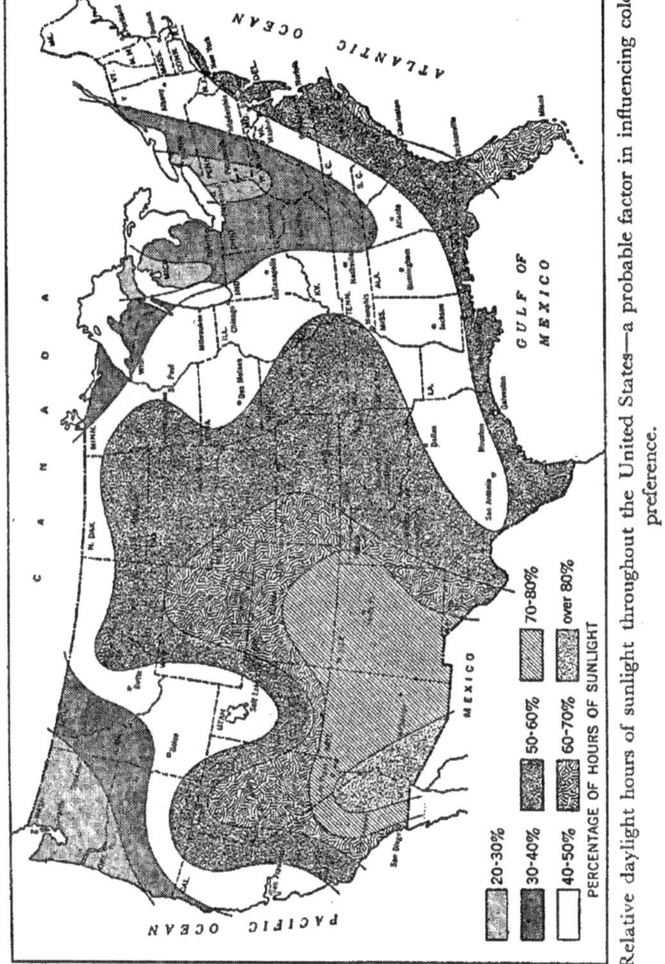

Relative daylight hours of sunlight throughout the United States—a probable factor in influencing color preference.

repeat, sunlight, rather than temperature, seems to be the influencing factor. The map opposite has actually been used by Mrs. Taylor to form a basis for the stocking of consumer goods. White shoes, for example, will sell (in relation to dark shoes) in about the same proportion in Boston as in Chicago, Nashville, Dallas, Butte, and Sacramento. This applies also to certain colors in men's neckties, knitting yarns, and hosts of other products.

Where sunlight is abundant, the colors wanted are strong, rich, and frequently brilliant, like red—whether in northern latitudes or southern.

Where sunlight is more scarce, the colors wanted are softer, duller, and a greater preference for blue is noted.

Those concerned with color trends and the vagaries of mass preferences for hue will perhaps find this map intriguing. It suggests a new viewpoint, a new approach to the study of regional color preferences. National organizations having branches in various parts of the country may well heed its implications. Sales territories are perhaps not to be drawn with a compass, and goods stocked as belonging to the North, the South, the East, or the West; for human likes may vary with the sun itself, the wife of the ranchman in North Dakota having much in common with her sister down around Waco, Texas.

Blue Preference of Adults

Man is for the most part a diurnal animal, but not altogether. While Nature has equipped him with eyes that see most effectively in daylight, she also perhaps anticipated that he might do some lounging in caves and huts. The human eye functions best under moderately bright light and cannot stand exceedingly high intensities, as can the eyes of some birds, the ground squirrel, and the prairie dog.

In many creatures nature provides eye droplets, colored corneas, and colored lenses (chiefly yellowish)—all to filter light and to aid visual acuity. Man in this regard is no exception. His eye, as well, has a certain amount of yellow pigmentation.

Psychological tests reveal an increase in preference for blue in adults. Is not this emotional reaction also to be explained by a physiological fact? Walls writes, "It has been known for many years that the adult human lens is yellow, but not until very recently has it transpired that this is actually of advantage to sharp vision in bright light." In the opinion of some, even more than this is to be explained.

The yellowing of the lens increases steadily through the years. The lens of a child may absorb 10 per cent of blue light; that of an old man, 85 per cent. Artists mellow with age frequently experience difficulty in the "handling of blues." It therefore may be safe to assume that another sort of accommodation takes place; the eyes of most of us are "thirsty" for more and more blue as the lens proceeds to filter it out.

As to the color purple, which seldom holds much appeal in consumer goods, here is a speculation by the author. Luckiesh in his book, *The Science of Seeing*, writes, "Inasmuch as it is impossible for the eye to focus blue and red light from a given object in the same plane, a purple object can never appear distinctly in focus." Maybe the idea is farfetched, but this confusion set up in the eye may have something to do with the generally low ranking of purple in sales records. This would again imply that human likes and dislikes have direct connections with things physiological.

Color and Character

In a later chapter of this book, a few rather extravagant comments will be found about color and personality. Now, however, let me append a few concluding notes on color preference as it is revealed in innate and racial character traits.

Athletes are said to prefer red, intellectuals blue, egotists yellow, while the convivial favor orange. Through national and religious traditions, some peoples—Hindus, Chinese—look upon yellow as a sacred and happy color. Mohammedans love green, the hue of Allah. The Irishman is prejudiced against orange, and many Scottish people dislike green.

The American, a blend of many nationalities, has all and none of these peculiarities. He will, of course, show patriotism to red and blue—and, happily, his flag combines the most universally pleasing of all hues.

All human beings, victims of their moods, fancies, and thoughts, will let brain and memory cause even further idiosyncrasies. The mere feel of some objects—an orange, a derby, a water glass—may conjure up an image of hue. A black-and-white photograph will promptly be visualized in color by anyone familiar with the objects in it.

In a similar way, it is human for a person to exaggerate color qualities when memory is called upon. When someone is asked to match a color, perhaps the blue of a person's eyes or the red of a person's hair, the colors selected will almost invariably be more vivid. This exaggeration works its way into speech. Witness such expressions as red as a beet, white as snow, pitch black.

Here are two more curious facts. When the human eye is asked to judge a color for its *hue*, a darker value will generally be picked. When the judgment is set for *value*, a person is likely to think of the color as being lighter than it actually is.

Chapter 4

THESE ARE THE THINGS THAT PEOPLE BUY

THE psychologist would declare that the hues best liked by mortals as a matter of choice of heart are blue, red, and green, in this order. His conclusions are interesting and to a large extent accurate, when checked in the light of sales experience.

Yet the purchase of many commodities involves other qualifying considerations that go beyond mere emotional pleasure alone. There are such things as vogues and trends, which may run contrary to tradition. A woman may rationalize over the problem of creating bright and airy conditions in a room. She may have common-sense theories about qualities in color that resist soiling, wear, fading, or that make her figure plumper or leaner. In such instances, she may temper her aesthetic notions somewhat with good intuition for the appropriate and the practical. Even here, however, she will usually follow the same conventions as her neighbor, thereby making a rule out of herself rather than an exception (fortunately).

For the present, we are concerned with consumer goods, with the styling of products and commodities themselves. We are not concerned with advertising, displays, packages—which are not the *actual thing* bought but rather the propaganda and the trumpet blowing, which are meant to call attention to it and to enhance it.

It is helpful, at this point, to understand that the product approach and the advertising approach do not necessarily involve the same principles and techniques.

1. The product approach may be termed an approach to beauty, a glorification and ennoblement of the things that

people buy, or the things that industry thinks they ought to buy. This is the commodity itself.

2. The approach to advertising, display, promotion is not so personal. Here the task is to stimulate action in the consumer—and to overlook whether or not she feels altogether happy about it. Good advertisements and good packages, for example, are assured through a practical knowledge of visibility and attention-value in color. The best principles to apply are scientific ones that have less to do with emotional factors than with those that are matter of fact.

In the first approach, you are part of the mob, a sort of business evangelist, struggling hard to give expression to human wants.

In the second approach, you are more of a sober doctor, diagnosing your patient with a sound knowledge of your profession and applying the needle or the pill to get the reaction you want.

Thus, in this and the next three chapters, the approach to beauty will be the one discussed. Following that, a number of technical facts about color and vision will be set forth.

Now, however, you are a student of people, and you are going to try to make sense and order out of their predilections.

Sales Records

It is never safe, of course, to do too much theorizing about color and people. Conclusions should follow evidence. Right colors sell in big volume; wrong colors do not. And right and wrong in merchandise are not unlike right and wrong in ethics—good, when the general interests of people are served, and bad, when they are not.

Anyone concerned with the problems of color in consumer goods will find real profit in a study of sales records. Such facts, however, do not make for very absorbing reading; yet, as they are pertinent to a book of this sort, some reference should be made to them. In Appendix *A* will be found a fairly comprehensive list of color preferences as they are revealed in actual reports from manufacturers. These figures, of course, are timely as of the writing of this book. While they naturally will undergo change, the reader may find some permanent interest in them as a source of future comparison.

The sales record holds something of the value of a stock market report. It bespeaks public demand, tells of peaks and valleys in the wants of consumers. Where such information is maintained over a period of time, the careful observer will soon begin to note the course of trends and to distinguish live issues from dead ones.

Colors, Today and Yesterday

Not so many years ago, the chief rage for color (in home furnishings) centered around green and rust. Dark colors were popular. Following this, there was a shift to deep blues and reds, and from there on up the ladder of the same hues to colors like periwinkle and rose. During the Second World War the typical American woman had a home in which the following taste was expressed.

Her house, if she had one, was colonial, painted white, with green roof and green shutters and trim.

For her living room (and for most other rooms) she used cream or ivory paint. On the floor was a rose or blue rug. The draperies had a cream ground with flowers in multicolor. The upholstery on her furniture was deep blue or wine (if she spent little money) or rose and medium blue (if she spent more money). Her dining room did not differ markedly from her living room.

In her bedroom, the wallpaper was pink, cream, or blue. Her blankets were rose, her bedspreads ecru or cream with rose or blue patterns. The covers on her mattresses were blue or rose. The walls of her bathroom were white (or peach, if she bought wall linoleum). Her towels were white or peach. Plumbing fixtures, if not white, were tan or ivory. Toilet paper, if not white or natural, was green.

In her kitchen, the linoleum on the floor was white with geometric designs in red and black. Two-thirds of the oilcloth she bought had red or green colors. In kitchen towels she preferred red as the best decorative color. In porcelain-enamel cooking utensils, she liked white with black, red, or green trim. In glassware, red predominated.

This does not imply much color harmony, perhaps, when judged in the light of sophisticated taste; yet it does reveal that, among many products, few colors were desired.

The American Market

To portray this story of color and people in its immensity, let us quote a few figures. In the last census (1940), there were in the United States some 131,600,000 human beings. Nearly 77 per cent were over fourteen years of age. They all lived in over 34,000,000 dwelling places—2.65 persons to each homeside. More than 25,600,000 of these dwellings had electricity, and 13,400,000 had telephones. Automobile registration exceeded 26,900,000, and ownership of radios was above 29,300,000.

Of the total population, about 90 per cent were native-born, and 10 per cent were foreign-born. There were about 29,000,000 housewives. Of the 52,000,000 or more persons willing or able to work, three-quarters were men and one-quarter were women. As to the particular tasks that kept them busy, 8,252,000 were operators of one type or another; 8,234,000 were farmers or farm laborers; 7,518,000 were engaged in clerical or sales work; 5,056,000 were experienced craftsmen or foremen; 3,749,000 were proprietors or managers; 3,345,000 did professional or semiprofessional work; 3,064,000 did miscellaneous labor; 2,111,000 were in domestic service.

As to where they all made their abode, 12.1 per cent lived in 5 cities having more than a million population; 16.7 per cent lived in 87 cities with a population from 100,000 to 1,000,000; 18.8 per cent lived in 985 cities with a population from 10,000 to 100,000; and 8.9 per cent lived in 2,487 towns having from 2,500 to 10,000 population. The urban population in the above groups numbered 74,423,000 (56.5 per cent of all), leaving some 57,245,000 (43.5 per cent of all) out in rural districts.

The people of the United States have, normally, a total annual income approaching 100 billion dollars. This is partially spent as follows:

Food	$16,865,000,000
Housing	9,506,000,000
Home operation	5,285,000,000
Clothing	5,261,000,000
Automobiles	3,781,000,000
Medical care	2,205,000,000

Recreation	1,643,000,000
Furnishings	1,422,000,000
Personal care	1,032,000,000
Tobacco	966,000,000
Non-auto transportation	884,000,000
Reading	551,000,000
Education	506,000,000

Retail sales amount to about 42 billion dollars in one year. This money is spent in outlets such as the following:

Food stores	$10,165,000,000
Filling stations and automobile stores	8,368,000,000
Department and dry goods	7,946,000,000
Furniture, household, and radio	1,734,000,000
Drug stores	1,563,000,000
Variety stores	976,000,000

Much color is involved in all these bulky figures; yet the patterns of taste follow consistent lines. In America, in particular, mass production and mass marketing have been successful not only because of human ingenuity, but because of mass social enthusiasm and the willingness of millions of souls—many of them racial strangers to each other—to glory in the same things.

Europeans who in their own native lands dress differently and have different homes will, upon coming to America, quickly adopt the wants of their neighbors. It may be that liberty and democracy are great merchandising stimulants—the forces that really make mass production possible.

What Causes Trends?

It is difficult to account for shifts in color preference. From the inside out, so to speak, consumers generally want a change of environment. Yet from the outside in, these changes are very difficult to direct. Merchandising failures exist by the score in which individual maufacturers, or groups of them, have gone in for extensive and expensive promotions, only to find that public fancy was obdurate indeed. A visit to close-out shops will tell a vivid story of colors that remained waifs and had no homes to welcome them. It seems as though people are always susceptible to the infection of color. They expose themselves to it always. When they are smitten, their fevers are likely to become something of an epidemic, so many of them contract the same virus.

Perhaps their emotional natures are sensitive in spots. You can try this and that, but nothing succeeds until their systems all of a sudden react. They cannot *think* themselves into color choices, any more, possibly, than they can think themselves into influenza (or out of it).

Yet color trends start somewhere. In women's fashions, Paris used to be an influence. Even more than the leading *couturiers*, however, public taste had been affected by the eminent personages who first exhibited the gowns. Some time ago, the Duchess of Windsor created a rage for tweed suits; and, in a past generation, Alice blue was made famous by a President's daughter.

In these days, the motion-picture star, the dilettante, the actress, and the model, the cocktail lounge, the important reception—these are the test pilots and the proving grounds where the unusual is paraded before the multitude and applauded or not.

Again, in fashions, one is not to forget the counterpart of the ancient courtesans—women devoted to the profession of voluptuousness and seduction, and usually supplied with the necessary cash to indulge it on a grand scale.

In more substantial products, such as home furnishings, the leading decorators and shops, the motion pictures, the colorful magazines must be credited with the origination of a number of vogues. Not that such vogues are established through well-laid plans, but that enough different things are shown until something "catches on." Once again, the consumer seems to be at her own mercy, like a box of wet matches. You strike one hue after another until, unexpectedly, one of them blazes and the conflagration is on.

Strange ventures in color may be profitable to magazine publishers. However, no manufacturer or store is advised to go through the same speculation. When you are in business to manufacture, merchandise, and sell commodities at a profit, you just don't mount the spectrum and ride off in all directions.

After all, it is impossible to indulge all hunches or fancies. Nor would such a policy be practical. People's likes may be something of a mystery; but you may always have the assurance that such likes are of a limited rather than an infinite order. And they generally concern multitudes of products.

An Approach to Research

Mass preferences for color usually fall into a restricted number of hue ranges. A manufacturer of one product, for example, seeing that the colors he sells best do not agree with those of another manufacturer, may conclude that his commodity (and, perhaps, a lot of other commodities) has something special about it. However, were he to see the sales experience of many dozens of products neatly charted in color and sample, the whole picture would be impressive for its rather surprising regularity. From such a picture he would be forced to conclude that the public at large has relatively few pet favorites—no matter what, how many, or how much it buys.

The sales experience of any one manufacturer or seller is significant—to himself and to others. Yet the more data assembled, the better. America is one vast market. Those who spend their money for paints or furniture, dresses or hats, are the same persons who buy a thousand other things. By studying their larger needs, one gets a pretty good understanding of them, not merely as they stand at a counter buying a lone item, but as they gather things together to furnish a home or a wardrobe.

As to trends, here are a few introductory notes to the next two chapters.

In many lines of merchandise, good selling colors fluctuate but slightly over the years.

Many of the changes that do take place involve a mere shift in order of preference, the best sellers always being favorites, regardless of other considerations. Red, for example, may gain ascendancy over blue or green, then drop back at another time.

Even where the trend may be more radical, the new thing will more than likely represent a modification or variation of the simple key hue. Rich colors may give way to pastel tints, to grayish tones, or to deep shades—but blue, red, green may still distinguish their appeal. People are loyal to their hearts' desires.

Chapter 5

GIVING THE PUBLIC WHAT IT WANTS

TO KNOW the color wants of the American public, to assure the right styling of consumer merchandise, to reduce waste in retail selling—these are and should be the credo of modern business. Yet, when an analysis is made of styling methods, both in manufacturing and in retail selling, the process of color selection is often found to be inadequate and crude. It fails to have that better scientific control so often given to other problems of production, management, and finance.

Although most manufacturers sell different types of products, the majority of such products reach the same buying public. Yet, business is not always conscious of the fact that, while the average consumer buys almost countless products having different *uses*, her ideas of color are far less complex.

Her average wardrobe of some 36 items will represent not more than five or six different color ranges.

Her average home, with an inventory of several hundred things, will contain not more than a dozen major color groups.

Surely, the job of selling today—or any other day—is to glorify human wants; and it is the responsibility of business management to know these wants with all possible certainty. For, whether or not color may have utility in the use of a product, it certainly will have important bearing on its sale.

Pertinent facts are necessary to styling. These facts must not only find their basis in sales records and market observations, but in a widespread survey and measurement of consumer trends. Business must understand the *many* wants of the consumer, in order to satisfy any one of them.

Unfortunately, the problem of color styling is not to be met through subjective and personal guesswork. The real solution involves a certain amount of hard work. Sales records will afford a good summary of past and existing wants. Market data will help to broaden the consumer picture. Information from dealers and salesmen will perhaps keep a store or an industry abreast of competitive efforts.

Yet the job is not fully complete until a final check is made with the consumer herself!

In the final analysis, it is the responsibility of the manufacturer to protect the retailer; and it is the responsibility of the retailer to see that the public gets what it wants. Their common ground is the consumer—and the better they know her, the better will be their success.

Styling Methods—Stores

It is enlightening to quote from a survey conducted by American Color Trends among some two hundred manufacturers and retail stores. And it is rather shocking to find that, although the consumer is the true "goddess" of merchandising, she seems to be the last person to be consulted in matters of color styling.

To deal with stores first, when the question was asked, "Have you been troubled with unsold stocks because of colors that the consumer would not readily accept?" 83 per cent of stores said *yes* emphatically. Only 9 per cent said no, while 8 per cent gave no answer.

There is waste here, and plenty of it. Obviously while *some* colors sell readily, others do not. One store reports, "We find that the same poor selling colors run straight through from apparel to yard goods and shoes and are always stickers in every department." There is good reason to wonder who decides upon such colors and how business goes about selecting them. It is apparent that better control is needed, better and more reliable estimates of consumer desires.

How does the retail store go about the purchase of colors for such things as fashions and home furnishings? Only 25 per cent of them maintain a color-styling department; and, of these few special departments, practically all act in an advisory capacity. (In the survey the central management of only two stores actually dictated over the color purchases of various departments.)

However, practically all stores hold meetings of buyers to decide upon color ranges. Many of them (40 per cent) hold such meetings four times a year, at the beginning of each season. Others hold meetings only twice a year (16 per cent), some weekly, and others at irregular intervals.

Of greatest significance, perhaps, are the facts and data gathered by the store and used as a basis for color selection. If 83 per cent of them have unsold stocks, then distress merchandise in retail selling today must rest with faulty methods of prognostication. Here are the sources consulted by most stores. (More than one source, of course, is used by the average retailer.)

Sources	Per Cent	Sources	Per Cent
Sales records	61	Outside professional counsel	25
Advice of manufacturer	58	Direct research with consumer	17
Advice of sales persons	52		

Would it not be logical to expect that the bottom item, 17 per cent, should be nearer the top, if not at the very top?

Sales records are undoubtedly helpful, but they perhaps too frequently tell the story of past rather than future wants. The advice of the manufacturer should be reliable, although it apparently is far from that. If the maker is no more thorough than the seller, then both are sure to make mistakes. One store offered this sound advice to the manufacturer: "Investigate consumer tastes and wants before foisting combinations which look well to an artist but are not readily accepted by the public."

On the matter of satisfying the customer, 41 per cent of stores state that color ranges as wide and generous as possible are offered, while 35 per cent adhere to restricted ranges (25 per cent had no set policy). Large inventories of color complicate the stock picture and may lead to distress prices. This is particularly true where many colors are sold merely for the sake of variety and without right attention to consumer likes.

As to color coordination, the survey has brought forth several interesting facts.

To begin with, 82 per cent of stores firmly believe that the colors people like in one type of product help to determine what colors they want in other products. Only 6 per cent think not, and 12 per cent have no answer. That is, what a woman buys in a hat may influence what she wants in a purse. The rug for her living room will have bearing on the materials she buys for her draperies.

Yet, when the question was asked, "Has color coordination proved successful in your experience?" only 27 per cent of the stores said yes; 46 per cent said, to some extent only; and 4 per cent definitely said no (23 per cent offered no opinion).

According to the survey, however, 76 per cent of stores have tried color coordination. No coordination has been attempted in 17 per cent of stores, while 8 per cent did not indicate a policy. Most coordination has been carried out, of course, in women's fashions and in home furnishings.

Here, then, is the enigma: If 82 per cent of stores say that the color of one product will influence the color of another, why do only 76 per cent of them attempt color coordination? And why do only 27 per cent say that color coordination is successful?

The answer must rest with the fact that, while matching colors may be pretty, they will not necessarily sell for this reason *in mass markets.* Until the consumer's wants are better understood and satisfied, color will continue to be a bugaboo, whether coordinated or not.

Styling Methods—Manufacturers

The manufacturers surveyed by American Color Trends sold a wide variety of products—building materials, home furnishings, wearing apparel, textiles, paints, floor coverings, wallpapers, etc. Over 84 per cent of them stated that color was of *major* importance, while 16 per cent considered it of average importance in the sale of their merchandise.

Asked if they have been troubled with unsold stocks because of colors that the consumer would not readily accept, 56 per cent said yes, 35 per cent said no, and 9 per cent had no comment. (To the same question, retail stores gave 83 per cent yes. Quite apparently, the manufacturer is able to sell to a store merchandise that the store, in turn, has real difficulty in selling! They both seem to make mistakes, although the store generally assumes the larger risks.)

Among the manufacturers surveyed, 58 per cent maintain color styling departments, while 42 per cent do not. Some 28 per cent of them subscribe to a list of 12 outside styling services; 72 per cent use no such service.

As to styling periods, 23 per cent of these concerns are most active in spring and fall, others at one period only, and still others throughout the year. New lines of merchandise are issued mainly in spring and fall, on January 1 and July 1, once a year only, or at any time during the year.

To assemble facts on which to base their color styling, manufacturers rely on five main sources (more than one source, of course, is used by the average company):

Sources	Per Cent	Sources	Per Cent
Sales records	76	Direct research with consumers	28
Advice from salesmen	66	Outside professional counsel	26
Advice from dealers	48		

It is strange that, although 56 per cent of them admit to unsold stocks on account of wrong colors, only half the number go to the consumer for research data. Asked if different sections of the country seem to demand different color ranges, 63 per cent said yes, 32 per cent said no, and 5 per cent gave no comment.

Few colored products stand alone in the consumer's mind. That is, she may make a hundred different purchases, each thing having a different utility; yet, in color effect, these products will all fall into a few restricted hue ranges. These are the colors for the manufacturer to determine and feature.

Thus, when asked if the colors people want in other products help to determine what people want in the manufacturer's particular item, 94 per cent say yes, while only 6 per cent say no. Despite this admission, only 37 per cent consider color coordination successful, while exactly the same number have found it only partially profitable. (Three per cent say that it hasn't worked at all, and 23 per cent make no comment.)

However, 58 per cent of manufacturers say they have attempted color coordination. On this matter, it is enlightening to compare the experience of manufacturers with that of the stores.

Color Coordination	Manufacturers, Per Cent	Stores, Per Cent
Tried	58	76
Successful (of those who tried)	37	27
Partially successful	37	46

Once again, the manufacturer has a better opinion of color (coordination) than the store. At least, the store more frankly admits the failure of coordinated merchandise to be profitable. While the *theory* is unquestionably sound; in *practice* it often fails. The reason for this must rest with the fact that too little effort is made to find out what the consumer wants.

Color coordination, of course, works better and is almost demanded in high-fashion markets. In mass markets there is some question as to its value. It is true perhaps that for every product purchased in ensemble, a hundred or more will be bought without reference to anything else. The first duty is to make the individual thing individually beautiful. If it can also be made to fit in with other things, so much the better.

How Many Colors?

The author holds the view that the aim of industry should be to produce as few colors as possible in any line of merchandise. This conviction is based on broad experience and extensive study of sales records. The argument is often presented that while few colors are desirable, an extensive range is necessary for prestige and to add "class" to a line.

This, it is probable, is mere assumption and hangs on thin threads. Salesmen will generally cry for more and more hues, and the production side of business will insist upon less and less. Even the more intelligent consumer will bewail the fact that so few colors are to be found, making it almost impossible for her to carry out some subtle scheme. The mass buyer, however, is never so finical.

Let us quote from the records of two lines of merchandise and discuss the economics of color ranges. The first of these manufacturers had 22 different colors, and he sold them in the following proportions:

Color	Per Cent	Color	Per Cent
Deep blue	24 5	Dark brown	2 0
Black	14.8	Dark green	1 3
Maroon	12.4	Vermilion	1 2
Navy	8.8	Grayish blue	1 0
Sand	8.2	Dark green	0 8
Medium blue	5.8	Silver	0.5
Red	4.3	Medium brown	0.4
White and natural	4.1	Yellow-green	0.3
Gold	3.7	Orange	0.3
Green	3.0	Pink	0.2
Light blue	2 2	Purple	0.2

This, according to the author's way of thinking, is a sad record. The spread from best to worst is entirely too great. For every one purple item he sold 123 in deep blue! Not to mention the probable complexity of the manufacturer's inventories, how much did the consumer have to pay *extra* for the privilege of looking at a wide variety of colors that she practically never bought? Was the salesman proud to have so much to offer the retailer? And if the retailer stocked up on most of these colors, was he proud to have them lying around the store unsold? (The colors in this particular line, incidentally, also came in different sizes!)

Now, consider the record of another manufacturer. While this is not the same product as the one above, it is not far removed and also involved the complexities of different sizes.

Color	Per Cent	Color	Per Cent
Black	21 6	Brown	10.5
Gray	15 7	Green	10 2
Blue	15 2	White	8.9
Red	11.1	Tan	6.8

Here the spread is far smoother, only three black items being sold to every one of tan. Probably the average manufacturer is not in a position to do quite so well, but he certainly ought to try.

In a recent experience of the author, a line of merchandise, previously sold in a range of 11 colors, was offered in only 7. The experiment was certainly modest enough, only four items being dropped from the line. The sales force, however, strenuously objected, making predictions that ominously pointed to disaster, claiming that dealers would reject the line altogether, and so on. However, the new seven colors had been carefully analyzed and tested. Each had every assurance of acceptance.

What happened is shown in the following data:

Color	Old Line, Per Cent	New Line, Per Cent
Dusty rose	17.3	25.0
Light blue	14.8	18.0
Peach	12.8	15.5
Deep rose	12.5	11.0
Green	10 4	13.5
Brown	7 5	
Maroon	6 6	
Beige	5.7	9 5
Yellow	4 9	
Deep blue	4.9	
Ivory	2 6	
White		7.5

The trepidations of the sales force were pretty well proved to be unfounded. First of all, the entire line sold much better, and far more total products were bought. The spread between best and worst seller was smoother—from 25 to 7.5 per cent, as against 17.3 to 2.6 per cent. Although most salesmen assume that dealers and consumers want many colors, the test proved the facts to be otherwise (as is often the case). The dealers, in fact, offered their blessings to the manufacturer for the advantages of a limited stock, ready acceptance, high rate of turnover, and almost total lack of unsold inventory.

How many colors should be offered? The question really ought to be, How few? To my way of thinking these are the factors that should decide.

1. A woman may haunt manufacturers and retailers for a strange color. If she is importunate, or if she happens to know the wife of the president, the hue may be stocked. One appeal, made by one woman, and insistently presented, may give the color a glory that is all out of proportion to its true value. What is the result? One (or a few) women are satisfied, and many are not. Inventories are complicated and distress

prices are nurtured—indirectly adding to the cost of the good selling items. Thus, will the loss of patronage (where a strange color is denied) be any worse than the costly waste of a slow-moving stock? This is a question to be weighed.

2. Far better is the policy of adhering to a few colors and making sure that these are exactly right, through intelligent research among consumers.

3. Obviously, the store and the manufacturer ought to know what people want. There is no possibility of catering to all finical taste anyhow. The only sure answer is to insist upon promise of fair volume, and to forego an abundance of colors. For your own assurance, however, know that the mass of people have simple taste and will not, as a rule, desire exotic things. It is your responsibility not to let a few voices convince you to offer something that a hundred other women will adjure.

4. If, for the sake of prestige or leadership, you want to be generous in your sales policy, take my advice and add more *designs*. A number of designs in a few good colors will sell. But a number of colors in few designs will not sell. This fact is too seldom appreciated; yet it is the one sure strategy for the company or store that wants to offer just a little more for the sake of being exclusive.

The Shape of Trends

It must be granted that people will not want the same colors, year in and year out. This observation, however, should not be too recklessly interpreted. I would rather say that people do not always want the same thing than to say that they *never* want the same thing, for the color trends of most products move rather slowly and clumsily in mass markets.

While trends are difficult to define and no less difficult to anticipate, a certain amount of reasonable order will be found in them. There are some products, nevertheless, that never seem to change in trend. Most of these are things that are bought chiefly on impulse. They range all the way from toy balloons to compacts; and the eternal values of bright red, blue, green, and white, best seem to complement them.

A general, allover picture of color trends will be found illustrated in an accompanying chart. This is based on an observation of buying trends over the years and was first brought to the attention of the author by an executive well experienced in mail-order selling and the merchandising of products in mass markets. Here are its generalizations.

Assume that, at a certain period, the colors most in demand are strong and rich (point 1 on the chart).

The next likely trend will be toward lighter tints or pastels, these colors being clean and clear in aspect. They will show steady rise (points 2, 3, 4) as the stronger hues begin to wane.

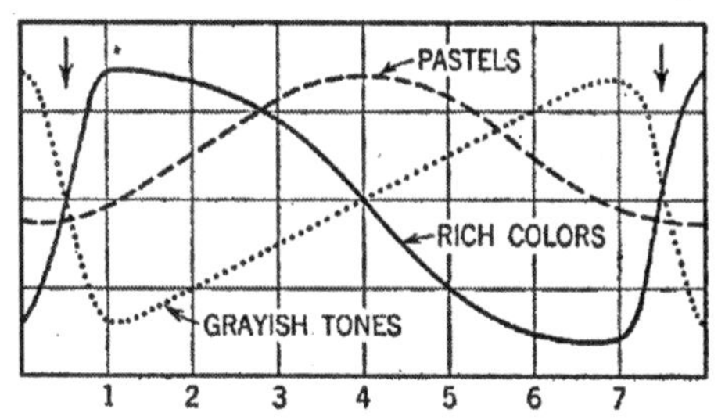

Color trends. Showing the usual course of color preference in many average lines of merchandise.

Following the pastels will be a trend toward grayish tones (so-called muted colors). Point 4 on the chart will be a very happy time indeed, with all forms of color selling fairly well —the pastels in the lead, the grays on the way in, and the strong colors on the way out.

With the ascendancy of the grayish, muted tones (points 6, 7), the strong and deep colors will reach a low ebb.

Then, more suddenly, the whole order will be reversed. Apparently the consumer will grow weary of this slow trek from richness to pastel to gray tone. Here taste, having become anemic, will rebel and long for strong tonic. She will become "smart" and "audacious," and her preferences will swing suddenly to vigorous colors (at the point indicated by arrows on the diagram).

However, while deep rich colors will have more sizable ups and downs, the pastels will trace a smoother course and will sell in more even volume.

As to the time duration involved in trends, the general movement is swift in women's apparel and slow in home furnishings, a complete cycle from richness to pastel to gray tone and back to richness again, requiring several years in most home commodities.

This is thinking broadly, of course. The interesting premise of the chart is that it pretty well analyzes the usual course of color preference and tells what form of color may logically be expected next. Beginning with the rich color, the consumer tends to find herself attracted to lighter modifications. Then this growth of appreciation leads to further subtlety, only to arrive at a point where the full beauty of color is sapped to an almost ashen grayness. A reaction takes place, and the cycle starts over again.

Economic and social conditions naturally affect trends. The depression of a few years back forced a quick return to pure color (possibly as a sort of psychological escape for the individual). Brilliance blossomed forth (in stoves, utensils, automobiles) where it had never been seen before. Since then, the trend has been clearly toward pastels and muted tones. The more ardent fever of the present day may likewise inspire boldness in color and give the cycle another whirl.

High Fashion

Catering to the whims of high fashion may be a complex and agonizing process, but it will pay big dividends to the "quality" business that exerts the right alertness. High-fashion buyers are likely to show unexpected fancies, and likely also to abandon them with the same alacrity. To sell exclusive products high in profit and low in volume, color coordination may be necessary. The manufacturer and the store may have to watch each other if they do not actually work together. All the influences of Fifth Avenue, the 35- and 50-cent magazines, the style shows will have to be attended. The gilded chariot of wealth (or presumed wealth) rattles along at a fast clip. You have to achieve the knack of jumping on and off with facility. While the rules are harder to write than for mass markets, research and a constant diagnosis of what goes on are both vital.

Sectional Preferences

Some data have already been given on color preferences in relation to hours of sunlight. For the most part, such differences concern either a liking for warm colors as against cool ones, or for rich chromas as against softer pastels or gray tones.

In this regard, the experience of one company manufacturing hard-surface floor coverings is significant. Generally, the two most popular colors are red and blue, the extreme regions of the spectrum. Further, people of Latin type—Spanish, Italian—prefer warm hues, while Scandinavian and Nordic types prefer cool hues.

New England is conservative. The preference here is for suppressed blues, reds and roses, greens, grays, tans—seldom the intense colors. Metropolitan New Yorkers delight in flashy contrasts. Exciting colors and color schemes of a sharp, clean nature are favored. New Yorkers, of course, run the gamut of the spectrum—from the highly styled colors of Fifth Avenue to the simple, crude palettes forever preferred by average mortals.

In the deep South (not including the Florida resort districts) red is the paragon—rich, flashy, burning. In the Southeastern states there is a bit of old-fashioned conservatism, though the favored colors are usually warm in character, like traditional Southern hospitality. Soft, medium tones, lacking in anything spectacular, are preferred.

In the Middle West and the Northwest, color choices seem to differ as the nationalities of settlers differ. City dwellers like brightness and contrast of clear, pure hues. Country dwellers buy more subdued tones of green, tan, and brown. And, again, Nordics will lean toward cool hues, and Latins toward warm. In Texas and the Southwest, hot climates (and perhaps Indians) inspire the consumer to want colors sufficiently strong in hue and contrast to vie with the intensity of the climate itself.

On the Pacific Coast, particularly in southern California, there is a real demand for light pastel tints—colors away from full intensity, yet clean and luminous in quality. Here even designs must be simpler. (Include the beach towns of Florida in this accounting.)

The Economics of Styling

Social values are becoming more and more important. In legislation and education, in public and industrial relations, a new attitude is being formed—an attitude that seeks to measure all human effort in terms of its practical significance to the masses at large. To fit in with all this, American manufacturers ought to recognize new responsibilities to the consumer. After all, the manufacturer has the duty of providing his dealers with colors that will sell; and the retailer must protect himself against loss and his customers against dissatisfaction. Distress merchandise is economically wrong and socially unfair. Goods bought at a bargain counter may seem cheap, yet the losses written off here must be charged elsewhere. Values are to be questioned when the maker—or the seller—must add to the price of his good items to compensate for his bad ones.

Business management, for the most part, has delegated matters of styling to certain employees, who generally have standing below that of a vice-president. The chief executive himself will usually confess to ignorance as far as "beauty" is concerned. He may handle finance, production, and labor; but he will turn to others for guidance in color. During the recent war period, however, management learned an important lesson. Governmental restrictions on color and the extent of color ranges demonstrated that considerable efficiency lies in such programs of simplification. It may be true that the consumer would buy almost anything because of scarcities. Nonetheless, styling became a function of management to the extent that an executive was usually called upon to dictate to his company what the government dictated to him. Many an organization with higher production, material, and labor costs—but with ceiling prices fixed on its commodities—made substantial profits, anyhow. Such profits came through the more efficient production of fewer items, reduced inventories, lower selling costs, and lusty consumer acceptance.

Perhaps styling, through this experience, will become more a concern of management. Certainly, the executive end of business should be able to speak with authority about color, or, at least, to set up means and devices by which it can check and double check the assumptions and vagaries of its designers. Such means and devices are to be found in intelligent research. They are a necessary function of management. They bring color into the front office and make it behave. They force it to keep regular hours, fill out an expense account, and conduct itself with perfect business demeanor.

Chapter 6

PRACTICAL RESEARCH TECHNIQUES

RESEARCH is indispensable to styling. As one successful manufacturer has stated, "You follow an artistic ideal and let me follow a statistical fact, and my products will outsell yours at least two to one." Not that there is anything wrong in a creative viewpoint, but that it lacks assurance of result unless it is confirmed by some tangible evidence.

Art is always an important consideration; yet it needs counsel and direction. For its own sake, however, it should be watched with a wary eye. Art in industry must have reason, sense, and purpose; and its expression must withstand the hard test of public acceptance. Fortunately, it is no more necessary for a businessman to experiment with color than it is for a doctor to experiment with a stomach-ache. Methods of diagnosis are available to both.

Basic Principles

To get at a basic technique, here are a few primary considerations. A fundamental plan is to be laid within the organization, and certain policies are to be established. All this is to be neatly integrated with direct research among consumers.

1. First of all, sales records should be maintained. These should include a company's own findings, as well as those of competition—if such facts are available.

2. If different products and different price ranges are involved, separate charts should be kept. Cheaper merchan-

dise will probably have a color acceptance differing from that of expensive merchandise.

3. Master charts are then needed, which will summarize the story for all types of color in all types of the manufacturer's products. These charts may be further broken down into market sections if the consumer shows singular preferences in various markets. Again, a differentiation should be made between low- and high-priced commodities. This effort will reveal a broad picture of consumer wants. (It must be remembered that color trends are of a general nature. Though the public may purchase an endless variety of things, it will have but a limited number of color preferences.)

4. If it is at all possible, the findings of the manufacturer should not be taken as conclusive until an actual check is made with his retailers, in order to determine if the public itself has accepted the goods. Unsold inventory lists may sometimes be conveniently examined by salesmen (or reports obtained through the mails). For example, promotions of individual colors by the manufacturer are often successful from *his* standpoint, but not from the standpoint of his dealers. Some time ago, a large retail chain made the rounds of New York close-out dress shops and found a staggering abundance of olive-green dresses. Although these dresses might have made a splendid showing on the records of the man who made them and sold them to certain outlets, they did not sell to the consumer—a fact that the manufacturer perhaps didn't even know. Though you may distribute your products through dealers, bear in mind that you are actually forever selling to consumers, even though indirectly.

5. Sales records and charts should be cumulative, and arranged by section of the country, if different habits are evident in different regions. The advantage here is to find out whether colors are on the way in or on the way out. With enough experience, saturation points will become known and may be recognized.

6. Finally, as will be emphasized later, the manufacturer should make an effort to acquaint himself with the sales records of related products. If he makes paint, it is important for him to know what goes on in rugs, wallpapers, draperies, upholstery fabrics. It is always vital to understand the

consumer as she buys many things. Color trends concern entire houses, rooms, wardrobes, more than they concern the individual items within these.

Deciding upon New Colors

Sales records form the basis of a profitable research program. They tell the story of the past. Factual information is constantly at hand for reference. The job now is to devise a method for predicting the future, for the styling of new colors, the development of new ranges and offerings. How is this to be done?

Consumer research is the one satisfactory answer. But before delving into the best principles to be followed, let us present a brief discussion of current theories and practices that ought to be skeptically questioned.

The personal opinions of artists and stylists (or of presidents' wives, secretaries, or stenographers) are sure to be unreliable, if not actually preposterous. They may, however, be temporarily kept in mind as ideas that will be later subjected to checking, analysis, and research estimate. Likewise, the advice of salesmen and dealers is likely to be prejudiced. Too frequently it is arbitrary and distorted, built up by argument or undue insistence, and thus supported more by exuberance than by fact. Again, however, such suggestions may be taken as ideas to be confirmed or disproved.

Consumer research as applied to color will shortly be explained in good detail. Here, however, it may be stated that no imposing amount of expense is going to be proposed. There are many singular qualities in color that make it a very recalcitrant subject to analyze. Notable among them is the fact that research on any one item is likely to be inconclusive. People's thoughts about color are seldom isolated. They are big thoughts. They must be studied as they represent a broad picture rather than a close-up of any particular detail within it.

In America, consumer research has achieved notable results. It has been employed in politics and business, by public-opinion agencies, advertising agencies, trade associations, publishers, and manufacturers, to measure the trend of human wants and beliefs and to form a basis for the intelligent satisfaction of them.

Methods of consumer research exist by the score: mail questionnaires, personal contacts on the street or from door to door, telephone interviews. It is important, of course, to get information from classes of individuals who are unmistakably interested in the product or in the question being researched. Here it is to be noted that persons of higher income and intelligence are more likely to be responsive; the masses are inclined to be taciturn. It is not always easy to get the right answer—from the right people. Yet, when a wise technique is followed, the results of research may be surprisingly accurate. In the field of politics, Gallup remarks that a poll of from 600 to 900 carefully chosen straw votes will often yield election results having an accuracy within 5 per cent.

Consumer Research—Negative

Egmont Arens writes, "I would like to warn against using people's opinions, having them tell you what they like, and then acting upon them. This is very dangerous. You are apt to be misled. The point is that what women *think* doesn't count; it is what they actually do under selling conditions."

This is entirely pertinent as applied to color, for its appeal is far more emotional than it is mental. Ways must be devised of sounding out predilections which are more or less innate; and these ways must avoid a number of pitfalls.

The method described below has been used with good success by the author and represents the policy of his research organization, American Color Trends. Frankly, the technique has been built out of much sad experience and failure; for, if business is not cautious, consumer research can lead to a collection of false conclusions and vapid assumptions.

In the first place, here are a few errors that should be avoided.

Color ranges are not well chosen by persons sitting around a conference table. No matter how many existing sales facts may be on hand, no matter how clear a new trend may seem, no matter how many years of experience may be back of the sales manager, stylist, or executive, conclusions should not be final without at least some research with the consumer.

Never think exclusively in terms of one individual product. If consumer testing is to be attempted, please know (from the hard-earned lesson of the author) that an isolated inquiry is likely to bring an equally isolated answer. A manufacturer recently offered a free sample of his product to women as a reward for giving judgment on a new series of hues. Quite to his amazement, the "abstract" choices failed to agree with the colors of the samples that the women wanted for themselves! And the result of the survey, incidentally, was far from satisfactory.

Never start "from scratch" in working out a line of colors to be researched, and never have too many. Walter B. Pitkin writes, "The ordinary man thinks fuzzily about his own wants beyond the commonplace. . . . The public knows its mind in general terms, never in detail." What I mean is this: if you begin from the beginning, so to speak, you are likely to do no more than get caught up with the past. Research must be projected, like a javelin, from the position of your last sales figures.

Consumer Research—Positive

The right formula, then—at least in the author's experience—may be directly stated in a number of well-ordered points. These are the principles utilized in the service performed by American Color Trends, chiefly for average consumer goods, such as home furnishings, sold in *mass markets*, not high fashion. (Supplementary data on women's furnishings will be given later in this chapter.)

First, sales figures are examined. Charts and statistics (cumulative, where such are available) are studied for the particular product under consideration. Similar data on competitive items are likewise noted.

The findings here are then checked against sales records of all types of merchandise, related and unrelated. We are striving to get at the larger picture of the consumer—at big trends, which carry little ones along with them.

At this point we observe what has been going on generally in the American market. With many facts before us, we know (*a*) what has been good and bad and (*b*) what color ranges have probably been missed. That is, have the records of other manufacturers of other types of products revealed good sellers that have not been offered in the item that we are considering? Ordinarily, what the consumer likes in one commodity will be liked in another.

To emphasize the above point, a fairly comprehensive picture of consumer wants often enables a manufacturer to break with tradition. Where his sales records may show good volume, for example, in certain types of color traditional to his business, a study of trends in other fields may expose different wants altogether, which he and his competitor may not have recognized. Such an instance is seen in the experience of a manufacturer of felt-base floor coverings. By tradition in this field, light backgrounds had been most acceptable in yard goods sold for use in kitchens and bathrooms. Rugs, however, were styled in deep tones of red, blue, green, tan. Objective research showed a certain "dusty rose" to be popular in a wide variety of other merchandise. On the logical deduction that dusty rose ranked high in the consumer's preference—and might be welcomed in a felt-base rug—a test was made and the rug was put on the market. Although it had no precedent (and no competition) in its own field, it immediately climbed up among the leading sellers and remained there for a long while. The manufacturer not only made an excellent profit, but higher prestige was realized among his dealers.

Assume now that we have a good knowledge as to what has been sold, and a fairly confident idea as to what might sell in the future. (At this stage we may consider the personal notions of stylists, salesmen, and dealers.) We are ready to submit our prognostications to the consumer! How shall we go about it?

Giving attention to as few colors as possible, we first take our leading sellers and prepare samples of them in other variations: that is, lighter, darker, grayer, or purer color. We keep in mind that color trends are likely (*a*) to rearrange the positions of the top sellers or (*b*) to present shifts in *form* rather than shifts in hue. We also keep in mind that we are trying to improve our good items as well as to groom unknowns.

Next, we study the best sellers in other types of products. Where we find a color or a color range that has been successful, we make an adaptation of it to our own commodity.

In the third place, we may include a few colors that the stylist or the salesman may insist upon, provided that his experience and judgment are tolerably sage.

What we have now are several potentials—a dozen, at the most—and there is good reason to be anxious about each one of them.

Going to the Consumer

From this point on, the research method for testing color differs radically from the method employed in polling such things as public opinion, advertising copy, the utility of a new product, or design and form preferences. The need for a unique approach is all too obvious: color in the individual product is irrelevant. For example, a woman might give a fairly sound opinion about the best size, shape, and convenient form for a kitchen mixer—and it would not be necessary to talk about anything other than kitchen mixers. But, if you were trying to decide upon the most pleasing color for the article in question, she perhaps would not know how to express herself unless you talked about kitchens themselves and the things that belong in a kitchen.

To speak of the method used by the author, major study and research are devoted to the over-all wants of the consumer—her living room, bedroom, bathroom; her spring or fall wardrobe; her basic concerns. This is where trends go on most actively, where the colors of individual articles are determined.

Hence, our visualizations consist of sketches or drawings in which ensembles are illustrated. We talk in generalities and ask general questions. We sound out the consumer's enthusiasm for color in home and dress. Some drawings show her the things that she already has. Other drawings show her a few variations of them. Still others (on the wise basis already described) show new departures.

We are not interested in "yes" and "no" answers. We are seeking reactions to new and general appeals, from which we can *interpret* the best color for any one or more products.

For our audience we go to three sources:

1. Women trained to lecture on color harmony before clubs and associations comprised of members who represent

our market. These women sound out the desires of the consumer when she is in the precise mood we want.

2. Special exhibits in retail stores and in those departments where our product is purchased.

3. Mail questionnaires, with a simple check list and a giveaway (on color harmony or character analysis) as an inducement. These questionnaires may be mailed to a selected list of people or delivered in person to organized consumer panels.

The facts so gathered, from as many regions of the country as we find necessary, are thus more general than specific. However, with the consumer's bigger wishes estimated, it then becomes fairly easy to seize upon the right colors for individual products. Now the deduction may be built upon good foundation and may further capitalize the particular sales records and experience of any one company. We have not cornered our answer so much as we have found the whole room in which it is automatically revealed. What we conclude upon is rather sure to be something that will fit happily and compatibly into its place in the consumer's household.

Finally, where it is practical to do so, and where time permits, we produce limited quantities of actual merchandise in color and rush them out, to see what happens under actual sales conditions.

Incidentally, we have not found the ringing of doorbells or the standing on busy streets to be practical. These types of research are unquestionably effective under many circumstances and for many purposes, but they do not seem to lend themselves to the problems of color.

The Benefits of Research

While the above method is fairly comprehensive and expensive, it by no means equals the high costs frequently spent for consumer polls. It is also more systematic and lends itself to year-in and year-out practice, as against a strenuous, one-time effort that has no permanent value.

In the service of American Color Trends we are able to consider the problems of a number of manufacturers at one time. Costs are thus low and equitably shared. This suggests like cooperation among groups of manufacturers, a policy that is sound because of the inclusive nature (as to products) of mass preferences for color.

Intelligent research pays big dividends. As one example, we may quote the sales and production records of a sizable company making a home product. In a normal year—before a factual styling program was introduced—70.2 per cent of the manufacturer's output was sold at regular prices, while 29.8 per cent sold as "off goods." When the colors had been vastly improved and better adjusted to mass demand through careful research, the 70.2 per cent figure rose to 80.6 per cent, and the "off-goods" figure dropped from 29.8 to 19.4 per cent.

Other things happened, besides. Because of the greater demand for individual patterns and colors, there was a 37.5 per cent increase in the average production run. This forced a drop in catalogued items from 135 to 107. Although the general price structure and the total volume of goods produced remained about the same, substantially higher profits were realized in quicker turnover, more efficient production, and reduced inventories.

The Anomalies of Fashion

The discussion of research in this chapter has been concerned mainly with such durable goods as home furnishings. The problem of women's fashions presents something of a mystery itself. The author confesses to some bewilderment over the vagaries of feminine taste in apparel. Hence, he has gone to others more qualified and experienced.

In a series of notes taken under the good counsel of Helen D. Taylor, much has been learned. In normal times, new fashions for color in women's apparel are inaugurated in the salons of Paris, New York, and—more lately—Hollywood. However, it is not so much the genius of the *couturier* that holds the eyes of the public as the eminent or conspicuous personages who go forth bedecked in the resulting creations. The taste exhibited by a princess, a motion-picture actress, the daughter or the wife of a fabulously rich capitalist—the taste shown by these rare beings who are the idols of the masses, will dictate the colors and the styles that represent the subsequent taste of the multitude.

In many instances, the color choices of the few elect are seized upon with surprising rapidity. At once, however, a process of elimination takes place. Not all of these exclusive colors, by any means, will find their way to the Main Streets of America.

In the styling of apparel for mass markets, the method appears to be simple enough. Expensive merchandise is closely watched. Fashion shows are attended. Checkups are made along Fifth Avenue. As certain colors appear to "catch on" and to move in satisfactory volume, tolerance—if not acceptance—is promptly shown. A few large organizations will add one or two such new items to a few of their higher priced garments—and wait still longer. The weeding-out process becomes almost automatic. Dozens of colors may be dyed up and shown about to dealers and consumers for an opinion. From among these, a few may eventually be considered reliable. Finally, a conservative choice is made, and out the colors go into vast production.

Even then, costly errors are frequently made. Strange colors are seized prematurely. There are also occasions in which the color has been offered too late, its saturation point in human desire having been reached.

Ordinarily, color trends in women's fashions sweep quickly across the country. Styles in conservative and street wear fan out from New York; and, lately, much fanning out has come from California in the realm of sportswear. Whereas the western regions of America used to be about a year behind Manhattan, now they crowd at its heels.

There is much cooperative effort in the styling of women's fashions. Through the able work of the Textile Color Card Association, for example, seasonal cards are issued. While such forecasts are not always followed to the shade, they do form a basic chart upon which manufacturers in many related and unrelated fields may work out their color ranges with some consistency and harmony. There are other independent groups, such as those in California, which get together and style their colors through mutual agreement and enthusiasm.

Although fashion colors rise unexpectedly and undergo radical change, particularly in expensive apparel, larger and more sober trends are to be observed in the reactions of the masses.

In the so-called Gay Nineties, fuchsia, purple, chartreuse, and brown were in greatest demand. The vogue of these colors slowly diminished, and they were replaced by blue, rose, navy, and black. Now and then, however, purple, yellow-green, and brown have their popular vogue, which fades out again before their rivals, blue and rose. Blue seems to be America's paragon of color, and articles of this coloring sell more steadily and more successfully than any others.

Of all the trends of these days, the one most obvious is that centering around a shift from dressy fashions to sportswear. House dresses and house shoes are rapidly giving way to more sporty types of garments. To complement the change, colors are shifting from more feminine variations (purples, lavenders, delicate pastels) to franker and more impulsive red, yellow, green, blue. Romance is going out and rationalism is coming in—perhaps.

If conclusions are to be drawn, it is rather safe to generalize that expensive apparel is subject to greater revolution, to more marked extremes, with good taste best expressed by soft, grayish, and refined colors. In inexpensive apparel, likes are more steady—with popular taste best expressed by the simplicity of primitive (not muted) colors.

Chapter 7

GLORIFYING HUMAN DESIRES

THE research of the psychologist into the mysteries of color harmony has been extremely limited and incomplete as compared with the broad experience and abundant case histories of industry. For every one inquiry on the part of science among a few hundred people, there are thousands of inquiries conducted in modern business among millions of people. Business, in fact, could teach science many a lesson, for its laboratory is the nation and its subject the vast multitude of population.

The problems of merchandising color may be summed up in these words: Find out what people like best, and then glorify these desires with all possible skill and effectiveness. To find out what people want you must study their natures and ask them many questions. To glorify color you must show artistic strategy that flatters less your own feelings than those of your audience.

While the management end of business will usually be quite open to advice (because it admits that it doesn't know much about beauty), the styling division will not always be so tolerant. People trained in aesthetics are inclined to have a subjective viewpoint, to claim some special talent that gives them an insight that nature has denied to most other mortals. This vanity is all right up to the point where it may lead a person to the delusion that his taste is the taste of the masses. The author knows the symptoms well. The products of industry are ugly, the designs wretched, the colors bad. Yet, when the creative spirit of an individual is let free, it very often fizzles out like a wet firecracker. Too much aesthetic energy leads to overstyling, and overstyling is bad—very bad—in selling to the masses.

Thus, to speak of beauty or color harmony is to take chances, and to give advice is to risk addressing many stopped ears; yet, the attempt will be made here. However, we shall try not to play the part of the theorist. The facts of this chapter are based on an extensive struggle with the sale of colored commodities and on an even more complete observation, analyzing the measurement of the successes and failures of others. This will be shop talk, not pedantry, and it will represent practical experience, not arbitrary opinion.

Color Systems

Color systems are well worth studying. They give the world of color a semblance of order. They make clear that colors, tints, shades, and tones all follow logical sequences and that color organization may be neat and not haphazard.

The two best known systems are those of Munsell and Ostwald. They will be briefly described in the last chapter of this book. It should be borne in mind, however, that the designers of color systems are not in the business of merchandising. Nor should they be, for they are more concerned with a graphic representation of the psychological order of color than they are with rugs or wallpapers. And even in their own province they are far from omnipotent. Newhall found the general shape of the Munsell solid far from right. The spacing of the colors is not well balanced, nor are the steps "equivalent perceptually," as they are supposed to be. Eysenck found Ostwald's gray scale and his spacing of pure hues both subject to criticism. He also submitted certain of Ostwald's theories of harmony to actual test and found them wanting.

With a color system a person is likely to be dogmatic, to let supposed order delude him into thinking that he has found some sure trick or formula to answer human demands. Some champions of Munsell might say, for example, that a color scheme that cancels into neutral gray (when corresponding areas of the colors are spun on a wheel) is a beautiful color scheme. This is mere assumption. Such a "law" is unfortunately too good to be true. Rules are profitable to understand if not to apply. A workman sorts his tools out on a bench. Yet, the mere order of his tools, while it may be handy and sensible, does not ensure his skill with them.

Color systems are not the answer to styling; they are merely an approach, a convenience for thought. Learn about them, by all means; but do not for the sake of orthodoxy win an argument and lose a sale. People are the legions you should try to command. Beware lest any cardboard forts of charts and gadgets make you vulnerable.

Single Colors

The problem, say, is to choose a single color for a product—a salt shaker, the handle of a knife, a small plastic radio. This one color is to sell in big volume and to be as essential to the item as are its size and shape. You may take an artistic attitude if you wish about what I have to say. You may or may not like the hues I would select; but my job is to reach the most people, and I would go about it in this fashion.

Where impulse counts (as in the array of things found in a 10-cent store) and where the products are not too large, the best colors naturally are blue, red, green, white (or cream).

The blue is clearly primary, neither greenish nor purplish in hue.

The red is a full scarlet, kept rather toward the vermilion side than the purple.

The green is clear and, preferably, not too yellow or too olive.

The cream or ivory is simple enough to decide upon. The white should be like "fresh driven snow," and free of any tinge of gray.

With pastel tints your best colors are sure to be pale blue (either of an ultramarine or of a turquoise cast, although the former is usually better), pink (on the orange side), peach or coral, pale green (not too yellowish), and pale yellow (like butter, and neither too much like lemon nor too much like goldenrod).

In deep shades, rich navy, maroon, warm brown, and dark green (not olive) will more than likely sell best.

Among grayish tones, soft blues and roses, tans, cool greens, beige, and taupe will be popular.

Much, of course, depends upon the product and upon the consumer's notion as to its most appropriate tone quality. However, general principles are to be observed in the following:

1. The consumer in mass markets will prefer the elementary color, regardless of its modification as a tint, shade, or tone.

2. She will not, as a rule, like in-between colors, purples, yellow-greens, blue-violets, red-violets, or even blue-greens. (Of the lot, blue-green will be the only one likely to have merit.)

3. As far as a complete line or range of colors is concerned, she will like few of them rather than many. In mass markets there is seldom any logic in offering a large number of hues. In practically every line of merchandise, whether it consists of 6 colors or 36 color, the top ranking 4 or 5 will frequently sell from 50 to 75 per cent of volume. To say that extra colors add prestige or enhance the leaders is not altogether true. A wormy apple hardly sells a sound red one by juxtaposition. Colors that people *do not* like may cause the rest to suffer! A wide selection of colors may confuse the consumer's mind and bother her decision rather than aid it.

4. In any range of colors, it is not always wise to adhere to one form of color variation. That is, you do not need all pastels, or all pure color, or all deep shades, no matter how pretty such balanced colors may look side by side. It is your privilege (and necessity, in many cases) to mix forms. A pale blue may sell better than a deep, pure blue; but a brilliant red may sell better than pink. And tan, which is neither a pure color nor a clear tint, may fit in almost anywhere.

5. It also goes without saying that a color that sells well in one form will not, by that grace, sell well in another. A big demand for red will not assure success for pink or maroon (though there may be some possibility that it will). It is necessary to watch your customers, to divorce yourself from purely arbitrary hunches, and to let the facts of actual demand speak for themselves.

Clarity of Form

There is one scientific principle, however, that helps to explain why clarity of color appeals to most people. Glance back at the Color Triangle illustrated in Chapter 2. In human sensation the three primary forms are pure color, white, and black. The secondary forms are gray, tint, shade, tone. Tints are whitish colors, shades are blackish, and tones are grayish. If it is part of the psychological make-up of people to see the world of color in this way, to sort all visual experiences into seven compact boxes, then a plausible deduction is to be made.

The principle, briefly, is that every color and every form should be lucid. It should have clarity and be a shining example of perfection in its particular classification.

Pure colors, if chosen, should be pure, brilliant, and as saturated as possible.

Whites should be unquestionably white and blacks, unquestionably black.

Grays, tints, shades, tones, when chosen, should all be unmistakable as such and *should not be confused with other forms.*

It should be possible to give scores of practical examples of the above principle. A few will suffice.

Unquestionably, the ugliness of a sheet of newsprint stock rests with the fact that it is neither white nor a distinguishable gray. It lacks clarity. "Tattle-tale" gray becomes an advertising theme to shock the sensitive natures of women over the chastity of white. Bleaches and chemicals are applied to white fabrics to make them as light as possible. Dingy whites do not sell; if, for some reason, the quality of "pure" white cannot be achieved in a certain product, the manufacturer would do well to abandon it and adopt something else entirely. One publisher dodged the issue of a cheaper paper stock by tinting it ivory and actually bragging about it.

Similarly, any color must shun the in-between stigma. A red, for example, slightly tinted with white may appear faded and weak, assuming an aspect that suggests shoddy materials. Yet, a red tinted with enough white to lift its classification out of purity and into the crisp charm of a pink may find its beauty restored.

In the case of orange, a slightly blackish cast may make it seem dirty and soiled. Enough black would shift it to brown, where, as a new form, it may again delight the eye.

While off-whites are something of an exception, even these require care. For the most part, clean bluish or pinkish glints are best. Any off-whites resembling an old tennis shoe left out in the sun and rain should be avoided.

Orderly Arrangements of Color

In appearance, the pure colors of the spectrum divide themselves into warm colors (red, orange, yellow), cool colors (green, blue, violet), and two colors of apparently neutral "temperature" (yellow-green and red-violet). While color circles exist by the score, not all agree one with another. Some complement red with green, others red with blue-green. Some work with three so-called primaries—red-yellow-blue; Ostwald uses red-yellow-green-blue at equidistant points; Munsell uses red-yellow-green-blue-purple.

Color theories and systems feature a number of rules that are too academic to warrant lengthy description. Harmonies are to be found in mathematically spaced intervals, in combinations of adjacents, opposites, split-complements, triads, tetrads, and so on. Although such rules may serve a good training purpose, they do not in themselves hold guarantees of beauty.

On this matter of harmony Guilford has again shown real sagacity. "There is some evidence that either very small or very large differences in hue give more pleasing results than do medium differences. This tendency is much stronger for women than for men."

Parry Moon and D. E. Spencer have come to a similar conclusion. ("Geometric Formulation of Classical Color Harmony," *Journal of the Optical Society of America*, January, 1944.) In orderly arrangements of color in value and chroma, pleasing combinations are found when the intervals between any two colors are "unambiguous." Refer to the chart shown herewith. This has been reproduced from the above-mentioned article. If the chosen color is yellow, for example, all other colors occupying the shaded area would be ambiguous (and unpleasant), and all other colors occupying the clear areas would be unambiguous, or pleasing. Thus, the two shaded sections immediately next to the yellow (chosen color) would be ambiguous, or unpleasant. Such hues would be an orange-yellow and a very greenish yellow. In the next two sections, which are clear on the chart, the element of pleasure would be evident. The hues in this region, an orange and a clear yellow-green, would be found unambiguous and therefore harmonious. Next, the two large shaded sections would present more ambiguous hues—red and green—which the average person would not consider harmonious with yellow. The rest of the chart, the "contrast" section, would show purples and blues, which are unambiguous and which would therefore be pleasing.

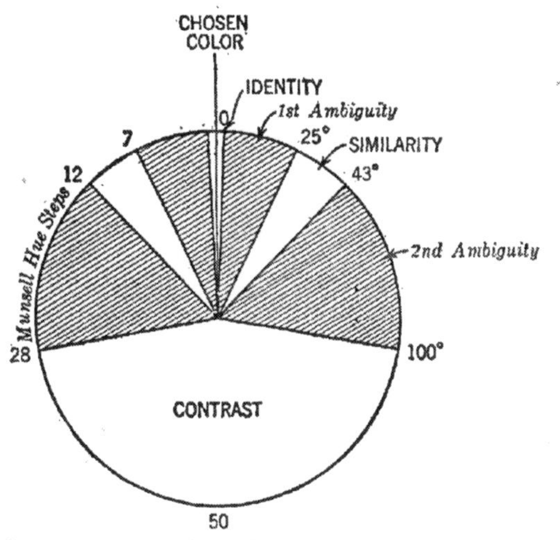

Ambiguous and unambiguous colors. For the chosen color above, the shaded areas indicate unpleasant harmonies, and the clear areas pleasant harmonies. (*After Moon.*)

The same general reactions—and the same general arrangement of the chart—would hold true for other hues chosen throughout the color circle. "Pleasing intervals and ambiguous intervals exist between colors, and an orderly geometric arrangement of color points leads to harmony."

A person, therefore, is likely to see harmony either in colors that are rather closely related, or in those which are antithetical and opposite—and not in other relationships. This conclusion thus suggests two big guiding principles—analogy and contrast. To these the author takes the liberty of adding a third principle—balance.

Analogy is an emotional thing, bringing together colors that bear close resemblance to each other and therefore give preference to one section of the color circle. For this reason, analogous schemes are either warm or cool in feeling. They are effective in the merchandising of numerous products—simply because any predominant color that the consumer may have in mind will be blended with its neighbors on the color circle and not compromised by opposite and, therefore, contradictory hues.

Contrast is more visual than emotional. It is particularly useful in advertising, packaging, and displays. Combinations of red with green or of red with blue hardly give any *one* of these colors a chance—particularly, if the areas are at all equal. Yet, contrast startles the eye, is full of visual thrill, and holds great attention-value.

Balance means a fuller diet of the spectrum. For example, combinations of red, yellow, green, blue include both analogy and contrast because of their completeness. Such balanced effects are highly salable and seem to gratify the instinctive love that people have for color in general.

For an additional taste of the technique of color arrangement, consider these points and observations.

1. All pure colors look well with white and black. This is something of a natural law of color harmony. In certain textiles and dress materials having multicolor patterns, in assortments of glassware or plastic, red, yellow, green, blue, presented separately or together, will have universal appeal. Such appeal will be particularly strong where impulse is a factor and the merchandise is relatively low in cost.

2. The same, however, may not be so true of intermediate colors, like orange, yellow-green, blue-green, purple. These latter do not have the arresting qualities of the psychological primaries. Yet, of assorted color schemes, if a combination of red, yellow, green, and blue on white is best, surely the second best combination is a tan, cream, or ivory ground embellished with orange or red-orange, yellowish green, brown, and (sometimes) a touch of purple.

3. Where a pure color is left predominant; that is, where it covers the major area of a product such as a cigarette case, a plastic radio, a set of toilet articles, other colors included for

incidental trim have to be added with caution. Often it is best to trim rich colors (and even pale ones) with white or black, gold or silver—nothing else. This leaves the main color undisturbed. A large area of red, for example, may be too vivid to be complemented with anything other than white or black; almost any other chromatic color may appear garish and discordant with it.

4. Where patterns are concerned, however, the above point may not necessarily apply. Then the observations of Guilford and of Moon and Spencer regarding analogy or contrast may be respected. While more advanced and subtle principles of beauty will be mentioned a few paragraphs farther on, let us include here a series of recommended combinations for pure hues—those suitable for a $1.95 house dress, inexpensive chinaware, oilcloth, and the like.

To glorify blue: For analogy on a white ground, use blue with black or navy. Or use different shades or tints of blue. (Only blue and green seem to lend themselves successfully to monochromatic tone variations. As a rule, pink is no help to red; and surely shades of yellow—which are olive—are no help to yellow.) For a blue background use white. For contrast on a white ground, use blue with red, but keep the red confined and the blue predominant. For balance on a white ground, use blue with red, yellow, and green in smaller amounts.

To glorify red: For analogy, use the red generously on a white ground and show touches or outlines of black. (But be sure that the red is a *good* one!) For a red ground use white, yellow, or black, or all of them together. For contrast on a white ground, use red with smaller touches of blue. (Red with blue sells better than with green or blue-green, regardless of the statements of color theories or systems.) For balance, use the red with yellow, green, and blue, leaving the red in the largest proportionate area.

To glorify green: For analogy on a white ground, use green with black. Or use different shades or tints of green on white. For a green ground have the designs in white or pale ivory, plus black, if desired. For contrast, on a white ground, use green with red (first choice) or with orange (second choice) or with purple (third choice). For balance, use large green areas on white with smaller areas of red, yellow, blue.

To glorify yellow: For analogy on a white ground, use large areas of yellow with black or deep brown outlines. Yellow, being a relatively pale color, may find most appeal when used as a background, to give it sufficient area. For contrast, use yellow with smaller amounts of blue. For balance, combine the yellow with its other psychological primaries—red, green, and blue.

Modified Colors

As to natural order in color harmony, I. H. Godlove has made an interesting observation. He writes, "If we note the lightnesses which are exhibited by mixtures of common pigments of 20 successive hues at their levels of greatest vividness, beginning with yellow and proceeding through green, blue, purple, red and back to yellow, we find the following well-marked sequence. The lightness goes down toward the greens, is lowest at purple-blue, and comes up again through red to yellow." This order for color (which has the form of an inverted pyramid when charted) Godlove terms the "natural sequence."

As far as average reaction is concerned, combinations of colors generally seem to be concordant and pleasing when the above sequence is respected—that is, when light tones of colors that in normal purity are light are combined with deep tones of colors that in normal purity are dark. "The indicated sequence . . . we may call the 'natural sequence,' as it occurs also in nature. Any sequence of hue and lightness opposite to the indicated one, for example, dark green-yellow and light blue, is an 'unnatural sequence.' It is found that practically all the harmonious pairs of colors are in natural sequence, and practically all the color pairs in unnatural sequence are bad."

To make Godlove's observation clear, in combining tints or tones the hue normally high in value may have the lighter tint, and the hue normally low in value may have the deeper tint. Here are a few examples.

Deep blue looks better with pale green than deep green looks with pale blue.

Orange buff looks better with deep violet than lavender looks with brown (which is a deep shade of orange).

Pink looks better with dark blue or purple than lavender looks with maroon.

Pale yellow looks better with brown or blue or violet than any pale greens, blues, or lavenders would look, for instance, with olive green (which is a deep shade of yellow).

There is further natural order to be found in the organization of the Color Triangle illustrated in Chapter 2. In the sequence of modified colors (tints, shades, and tones), any straight path on this Triangle leads to a concordant beauty.

First of all, as has been previously mentioned, all pure colors look well with white and black—these combining the three primary forms.

Combinations of white, gray, and black harmonize, with any of the forms satisfactory as a background.

Combinations of pure color, tint, white harmonize, with white best for background.

Combinations of pure color, shade, black harmonize, with black best for background.

Harmony is also indicated in diagonal paths:

Tints harmonize with tones and black. The best sequence is a black background, with tones for the pattern, and with the tints confined within the tones.

Shades harmonize with tones and white. (This arrangement, however, is less pleasant than the others.) The shade form would perhaps be best for background, with the pattern in tones and with glints of white within the pattern.

Pure colors harmonize with tones and gray. In this harmony, the effect is best when the ground is gray and the pure colors are contained inside the tones, for neat visual sequence.

It must be stated, however, that any remarks about the Color Triangle are more on the theoretical than on the practical side. The author does not advise its use in styling, other than to suggest its interest as a remarkable graph of the natural, psychological order of color. At the same time, there are good reasons for the harmonies of the Triangle—and some of them may well be remembered. In any straight path on the chart, the forms have elements in common. The sequence is always natural and orderly, and the eye is quick to sense pleasing relationships.

Again, a tone, which combined all three primary forms, also naturally blends with them. In fact, a tone (not gray) is the most neutral of all color forms. A purple tone, specifically, will blend with a wider array of colors than anything else, with the exception of black—for purple is neutral in "temperature" as well, bringing together as it does the two extremes of the spectrum.

Color Scheme or Color Effect?

The art of color reaches grander heights of expression than those so far described. For enlightenment on the finer points and intricacies of harmony the reader is referred to another book by the author, *Monument to Color*, which explores certain complex but highly effective principles derived largely from recent inquiries in the science of psychology.

In textiles and rugs, for example, real ability is required to create distinguished and salable works of art; yet, success in color requires as much resourcefulness in studying people as it does careful training and talent.

In high-fashion markets it is known that softness and refinement are wanted. Neutral beige, smoky blue, and rose may have ready acceptance as ground colors. Color harmony with these muted tones may demand plenty of skill. However, the appetite of more sophisticated taste must never be accepted as a criterion of mass taste. For a dowager, suitable neutrality in the basic decoration of a room may mean taupe blended with soft pastels. But for the wife of a laborer, neutrality may be quite differently interpreted to mean white bedecked with red, yellow, green, blue; for to her this full array of hue may more clearly suggest harmony with other things.

As a matter of viewpoint in selling color in *any* market, it must be recognized that people are essentially attracted to color *effects* rather than to color schemes. (The one exception is the balanced combination involving the use of red, yellow, green, blue on some off-neutral or white background.) *In the main, human preference for color clearly reveals a love for the individual qualities of individual colors. Color harmony in industry, therefore, is a problem of correlating all elements in terms of one predominant hue. It is not a color scheme but a color effect that is demanded*

—such effect to glorify and enhance the one preference that the buyer has in mind and heart.

A textile or a rug, therefore, is best sold as predominantly rose, blue, beige, and the like. This may seem elemental. Yet the idea of a color *scheme* persistently distracts the stylist from a more compelling approach. Why not express it in terms of a color *effect?* Let us explain the difference.

Assume that red is a good selling color in inexpensive floral carpeting. Assume, also, that this fact is known throughout the industry, all competition offering red on the market.

How will a better product, a better red be designed? Pattern will count, of course, but not too much. Will the stylist go from red to maroon or mulberry, to rose or pink? If he does, he will take chances; for, if simple red is what the consumer demands, he may make the mistake of digressing from it and lose volume accordingly.

If he is loyal to red, however, he may be very likely to try to harmonize it with gold, green, blue, and what not, playing with the spectrum as though he were assembling pieces in a jigsaw puzzle. He may be unduly concerned over color *schemes.*

There are, however, other strategies for him. These strategies relate to color *effect.* It is possible in the conception and blending of colors to get a wide variety of different appearances for the same hue. There are ways of making a color appear lustrous, iridescent, luminous—not through any chemical formulation of pigments and materials, but simply through the illusions created by certain color relationships.

To be somewhat technical, relationships involving a uniform suppression of various colors in terms of black will make a pure hue seem lustrous by comparison.

Relationships involving a uniform suppression of various colors in terms of gray will make a pure color seem iridescent.

Relationships in which all incidental colors seem to be pervaded by a key hue will make the effect luminous.

(The reader is again referred to *Monument to Color* for details and techniques.)

Now the stylist *has* something. He will create singular and magnificent products, which surpass everyday conventions and precedents. He will give his competitors much to worry about and, incidentally, to copy. His red, though it may exactly match the red of other companies, will have a unique and vital appearance through the startling relationship that it bears to the other hue elements in the design. It will have a different visual quality, a different effect! And it should put competitive merchandise to shame.

Keeping on the Bright Side

There is one final suggestion, and it gets at another bad habit in styling with color. Psychological investigation has shown that, where purities differ, people will like the richer color better; where values differ, the lighter tint will be preferred. It is inherent in people to see color (and to like it) for its clearest forms and expressions.

Those who work with color, however, are inclined to grow weary of the obvious. Artistic temperament and plain ennui will lead many a person to give honest colors "a touch of this or that," and to shun the frank quality for the elusive one. This may frequently prove all right in high-fashion markets, where the anomalous may be preferred to the conventional. But in mass markets, if there is ever any question as to what particular tone or quality a color should have, keep it on the pure side or on the bright side and you will be acting with sage wisdom.

When you weaken or otherwise modify a presumably monotonous color, to spare it from the commonplace, to make it exclusive, you may for your pains push the masses of your customers farther and farther out of reach. Far better will it be for you to take a deep breath, make the color obvious, spontaneous, and even vulgar (your own feelings aside), and thus be surer of getting your public to move along in your direction.

Chapter 8

THE HUMAN NATURE OF VISION

IN THE styling of consumer goods, a person ought to know as much as possible about the likes and dislikes of people. In many other applications of color in business, advertising, and merchandising, however, the approach may be less personal and directed toward a fairly scientific exploitation of known laws of vision. To sell a woman a rug or a dress, you must find out her choice of heart and then flatter it. But for compelling her notice, startling her eye, moving her to action in advertising, packaging, displays, it is perhaps more effective to work deliberately with the spectrum and get it to perform as you see fit. The human touch may thus give way to a well-ordered science that gets its results through command rather than ingratiation.

In styling, you intrigue the volition of people. In advertising, you give your audience a good shaking to arouse it from indifference. In the first instance, you may offer color on a silver platter. In the latter instance, you may not be so solicitous and may be more inclined to use the platter either to shine light into the eyes of people or to give them a gentle rap or two over the head.

To lead into such matters as attention-value, legibility, visibility, there is interest as well as value in reviewing a few of the singular phenomena of vision. Vision is, of course, the most fascinating and intricate of the senses. The eye is far from a camera in action and function, as it is intimately tied in with subtle mental processes. The perception of color is not a matter-of-fact reaction to light waves. On the contrary, it involves complex mental interpretations that are very astonishing to consider.

Types of Colors

Why is the human eye sensitive to only one particular region of the electromagnetic spectrum? Why doesn't it see infrared or ultraviolet? Walls sets up an explanation that the vertebrate eye had its origin in water. "It is thus no mere coincidence that the visible spectrum is roughly the transmission of water. The rod spectrum is closely fitted to water, the cone spectrum a little better fitted to air."

Again, there is a curious independence between color as sensation and color as radiant energy. It is human, for example, to see colors as filmy and spatial, like the sky; as three dimensional, like smoke or a bottle of liquid; as opaque and structural, like a painted surface. In principle, at least, three blues might thus be developed, all identical as to wave-length composition, but each with a novel, psychological appearance.

Further independence of vision is to be noted in the fact that, while colors may be fashioned in many ways, the eye often tends to give them likeness. A sodium vapor lamp transmits yellow rays only. An amber incandescent bulb, on the other hand, may transmit red, and green in addition to yellow—yet it may look the same as sodium vapor. All colors are not of a so-called chemical nature (paint, dye). The rainbow, the luster of an opal, a drop of oil on water, iridescence in a peacock feather—these are phenomena of light caused by refraction, diffraction, interference, polarization, etc., and are more related to physics than to chemistry. No pigments may be involved. The legerdemain is such that light is scattered, split apart or otherwise broken up into different hue elements. All these colors are described as red, green, blue, and the like, regardless of their physical nature.

Guilford writes, "It is a conviction of those who work with colors, that a certain observed color depends upon identical brain processes, no matter what stimulus initiated it. For example, the same green can be initiated by a certain monochromatic light or by mixtures of other wave lengths. So long as the observer cannot discriminate at all between the various greens thus produced, his brain processes are the same. From this one can see that the relationship between affective value and hue is a more important problem than the relationship between affective value and wave length."

In the realm of illusions, the brain may trick the eye, and the eye may contradict external fact. Black, while zero or nearly zero as far as physical energy is concerned, is quite positive as a sensation. Helmholtz, a great physicist, wrote, "Black is a real sensation, even if it is produced by entire

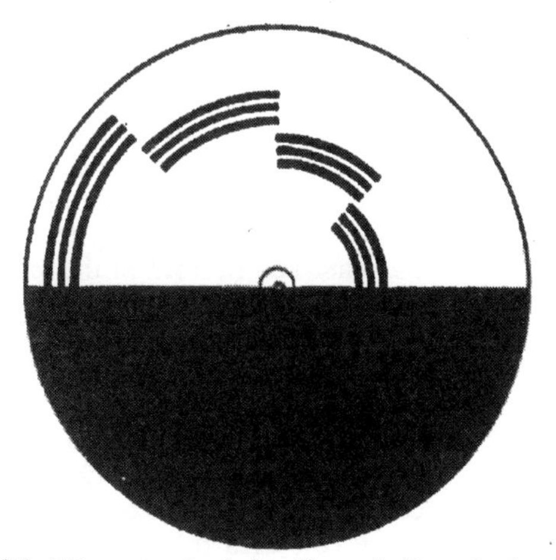

Benham's disk. When spun slowly it will reveal glints of red, yellow, green, blue.

absence of light. The sensation of black is distinctly different from the lack of all sensation." Black, therefore, is a color, and it will change the aspect of all colors it influences.

All this means that the eye (and the brain) pays but imperfect heed to the physics of the situation. Grays mixed through a combination of black and white may not be different in sensation from grays formed with red and green, or orange and blue. An area of yellow will appear brighter on black than it will on white. Pressure on the eyeball will cause sensations of color. The visualization of an area of red may bring up an afterimage of its complement (green). Cotton and silk, both white, may appear different in the quality of luster.

Queer, indeed, is that situation in which chromatic colors are experienced from black and white areas alone. Make a copy of Benham's disk shown in an accompanying illustration, using black ink on white cardboard. When it is spun slowly to the right, the inner rings will be dull red and the outer rings dull blue. When it is spun to the left, the order will be reversed.

The Eye

If in advertising, packages, and displays, color is to serve effective ends, then a review of some of the curiosities of the human seeing machine may well be given. While the eye is somewhat like a camera, the comparison is not to be pushed too far. Over the eyeball is the cornea, a transparent outer covering shaped like a watch crystal. Behind this is the pupil, an opening through which light enters. Within the pupil is the lens, capable of muscular adjustment to focus near and far objects. The iris, around and in front of the lens, expands or contracts to regulate the size of the pupil opening—wide in dim light, narrow in brilliant light. Back of this is the retina, a network of sensitive nerve endings, where the impulses of light are transmitted to the brain.

The retina has two regions and two functions. In the center is the yellow spot (macula) with a central pit (fovea) where the so-called cones are concentrated. These cones do most of the seeing job; for only here (the area measures about 1/16 inch across) does the eye see color, fine detail, surface structure, luster, form, and most of the perceptions associated with vision.

On the outer boundaries of the retina is the periphery, populated chiefly by rods. These nerve endings are more primitive. They are extremely sensitive to light, but not to color or detail. They detect motion but not the subtleties of form.

Vision, thus, has a sort of double action. Most of the job of seeing in bright light is accomplished by the fovea. As illumination decreases, the cones grow dormant, and the task of seeing goes over to the rods. Now (in twilight) the world is colorless, rather flat, and without evident detail. Similarly, the foveal and peripheral regions of the eye may work together as helpful partners. While the boxer may assign foveal vision to the chin of his adversary, his peripheral vision may keep him alert to haymakers swinging in around from the sides. Animated displays, signs, and posters in bright contrast may likewise catch the attention of a person very soberly making his way through traffic with eyes front.

In its sensitivity to brightness, the eye is astoundingly versatile. From sunlight to the feeble glow of a candle, human vision responds effectively. The range, in fact, may be in the order of a million to one. The eye also works quickly, registering impressions of objects and colors that may be exposed to it for merely a fraction of a second.

Visual Acuity

The eye has different sensitivity under different conditions of illumination. In the light-adapted eye, the spectrum is brightest at yellow and yellow-green. In the dark-adapted eye, the spectrum is brightest at blue-green, and red may fade out into blackness. In comparable fashion, the eye will have different acuity when its field of view is illuminated by lights of different hue.

While this so-called spectral quality of light puts many lighting engineers at loggerheads, the major facts are to be rather safely stated. Generally speaking, the eye sees well in natural light equivalent to daylight. Acuity is reduced as the spectral quality approaches blue. It is substantially high (perhaps even increased) as the spectral quality approaches yellow and orange—but not so far as red.

Some claims have been presented for mercury light (decidedly greenish and bluish). It is said to increase the apparent distinctness of certain objects, an observation that perhaps has some merit of truth. However, a yellowish quality in a light source is far easier to explain and defend from the standpoint of acuity. Luckiesh points out that yellow is the region of maximum selectivity, the brightest portion of the spectrum. It is without aberration (that is, the eye normally focuses it perfectly), and it is psychologically pleasing. By experiment, Luckiesh also demonstrated that by filtering out blue and violet radiation in a mercury light (also, in a tungsten lamp) visual acuity remained practically constant, despite the reduced amount of light absorbed by the filter. This would mean that, as far as visual acuity is concerned, yellow has definite advantages. Sodium light, for example, is highly efficient, although its distortion of colors makes it impossible for use under many circumstances.

Ferree and Rand placed yellow illumination at the top of the list, orange-yellow second, followed by yellow-green and green. Deep red, blue, and violet were least desirable. Blue, in fact, is very difficult for the eye to focus and will cause objects to appear blurred and surrounded by halos.

Under extreme dark adaptation, however, the eye seems to have best acuity under red light. Red illumination has been widely used for instrument panels in airplanes, for control rooms on ships and submarines. It has little influence on the dark-adapted eye and is not, in fact, seen on the outer boundaries of the retina, where the cones are lacking. It is therefore suitable as a blackout illuminant and will fail to stimulate the eye except when its rays strike near the fovea.

These facts have many practical applications. What are the best colors for sunglasses? What tint of paper is easiest on the eyes? What is the best quality of light for displaying merchandise?

In the case of sunglasses, yellow and yellow-green are unquestionably best. These colors may not only increase visibility and acuity by cutting down the excess brilliance of full sunlight (which may overtax the retina) but they screen out the disturbing influence of ultraviolet. In fact, on a sunny day yellow glasses may actually improve vision and help the eye to pierce distance. On the other hand, however, and more for the sake of illusion and fancy, blue glass may "cool" the sunlit world and give the eyes a glamorous appearance. Consumers, incidentally, do not favor the idea of yellow-green, regardless of its scientific advantages. Far more are they taken by the proverbial rose-colored tint. And as sunglasses are designed essentially to reduce glare, the tint may well be at once flattering and functional (amber, pinkish, or bluish) to the satisfaction of most vanity.

As to paper stocks for books and ledgers, there is no evidence that the universal use of white is wrong. Here, contrast is necessary for good visibility. Any colors that reduce this contrast usually handicap the eye. Shiny papers are, of course, unwanted. However, colors other than white would, for the reasons given above, be limited to soft cream or ivory tones—certainly not including greens or blues.

In the store interior and display, the simulation of natural daylight is too frequently misconstrued. It should be realized, for example, that the spectral quality of sunlight may vary more from noon to dusk than from "daylight" filters to ordinary tungsten lamps. Because of a phenomenon termed color constancy (to be described later), the eye has a fairly good sense of color values in any light that is halfway normal and not too monochromatic. Certainly, to use filters or bluish bulbs is a waste of money for large installations—such filters screening out from 10 to 70 per cent of light energy. For the same investment it would be better to have, say, 40 foot-candles of yellowish light than 20 foot-candles of "daylight." Extra brightness here would far outweigh any advantage of dim "natural" daylight.

Nor should psychological factors be ignored. That an illumination approaching natural daylight is better than any other is more assumption than fact. When science and illuminating engineers go to all the trouble of matching daylight, the public very often shows a dislike for it. So-called daylight fluorescent fixtures may be installed in a woman's kitchen, but seldom in her living room. Many stores and restaurants have found that "daylight" illumination actually repels trade and shows many types of merchandise and foods to disadvantage. People like a mellow "orangish" quality in light—for psychological reasons, perhaps, and even though it may cause distortions in the colors of merchandise. Daylight may be cold and forbidding.

Today, however, combinations of mercury with tungsten, or fluorescence with tungsten, often achieve good results economically. The fluorescent light, by itself, must be watched, for its spectral quality may be somewhat queer. The "daylight" tube is inclined to be bluish and deficient in red. The "3500°" tube emphasizes yellow-green, yellow, and orange. The "soft-white" favors the orange and red portions of the spectrum. Under most circumstances the bluish tint should be avoided. It may give a daylight appearance to merchandise, but it may also make the complexion of the buyer seem cadaverous, should she glance at herself in a mirror. Keep on the warm side! Such a strategy will not only assure a mellow and friendly blush, but it may also minimize ultraviolet radiation, which (in fluorescent tubes) may fade your merchandise.

Visibility

The study of light signals has revealed much about the carrying power of color. Granted that the brighter the source, the easier and farther it will be seen, red is perhaps the best of signaling colors: It is easily produced, instantly recognized, and plainly visible even at low intensities. Green is second, yellow third, and white fourth. Blue and purple are both difficult to see and to distinguish. In extremely dim light, however, when the eye is fully dark-adapted, blue will be seen over a wider range of vision than will red. In fact, the rods of the eye, most active in dim light, may not see red at all. (The official blackout illuminant during the war was a soft orange. Instrument panels on airplanes, for night flying, were illuminated by red light.) These facts may seem contradictory. In most conditions of merchandising, however, dark-adapted eyes just do not exist and are, hence, of little concern. Therefore, red in a neon sign is one of the very best to use—quick to attract the eye, quick to be recognized, and easy to read.

Legibility

When all colors are seen in the same light, however, the hue of highest visibility is yellow. Yellow, in a word, is the brightest and most luminous of colors. This accounts for the fact that the most legible of all color combinations is black on yellow. Next in order are green on white, red on white, blue on white, white on blue, with black on white sixth, and with the weaker combinations being red on yellow, green on red, and red on green. (Combinations of red and blue are too hopeless to include.)

It must be kept in mind, however, that, while white makes a good background, it lacks attention value. There is a difference between ease of seeing and eagerness to see. White or pale ivory may be ideal for a book; but for a package, poster, or display, the more exciting fascination of pure color seems demanded.

One thing to remember is that the eye wearies of color if exposed too much and too long to it. Black on yellow makes an excellent poster. Yet in a book the yellow would soon grow trying. It would build up a strong afterimage of blue, and it might also be emotionally monotonous, if not displeasing.

Attention-value

Red and yellow are the best attention getters, the one being aggressive, the other extremely bright and visible. They hold a never-failing attraction to human eyes. The question might be raised, however, as to their relative appeal and beauty. Let me caution the reader that a fairly tricky matter is involved here. The colors that people like best are not always the colors to which they will show the most active response. Again, one is reminded of the difference in "philosophy" between designing colors for merchandise and designing colors for posters, packages, and advertisements.

A few years ago, the Wrenn Paper Company released data on a series of tests with color. Blotters in different hues were delivered to several hundred people, who were later questioned. Here are three results.

1. When asked if they recalled having received a colored blotter (any hue at all), 61 per cent of women and 69 per cent of men said yes. Women who had received dark blue, olive green, purple, yellow, and red blotters did a better job of remembering than did those women who had received pale blue, violet, dark green, orange, and light green blotters. Among men, the greatest stimulation to memory was found in violet and dark blue, the least in light blue and red.

2. In testing the ability of people to *name* the color received, dark blue, light blue, and violet stood highest for women, and violet, light blue, and purple for men.

3. As to reader interest, the best colors for women were dark blue, light blue, and violet; and the best colors for men were violet, dark blue, and olive green.

Such research needs plenty of interpretation before it is too broadly accepted. "Stimulation to memory" does not mean either ease of recognition or power of commanding attention.

Colors that have novelty do not, by that grace, hold the most attraction to the eye. Posters and packages, in other words, may still be red or yellow, regardless of their banality; for red and yellow will win more glances than any other hues.

Again, it depends on whether or not you have attention to begin with. If you do (say, in opening an envelope), novelty may make more of an impression than conventionality. But if your task is to distract eyes toward signboards or containers that stand against a maze of competitive interests, then the primitive qualities of red and yellow may best help you attain your objective, regardless of what the public *thinks* about the matter.

In merchandise, in many types of advertisements, in illustrations, booklets, catalogues—where your product or message actually rests in the hands of your buyer or prospect and has his full attention—it may be all right to be somewhat aesthetic in your attitude. However, in a poster or a package—where you must shock people out of indifference—you would be wiser to take more of a scientific attitude and to pay rather close attention to the facts given in this chapter.

Dimension

Because the focus of the human eye is not the same for all hues, the colors of the spectrum will appear near or far, large or small, in consequence. Red, for example, focuses normally at a point behind the retina. To see it clearly, the lens of the eye grows fat (convex), pulling the color nearer and thus giving it apparently larger size. Conversely, blue is focused normally at a point in front of the retina, causing the lens to flatten out and push the color back. Walls writes, "Since the dioptric [refractive] apparatus ordinarily places the yellow focus in the visual-cell layer, we must actually *accommodate* when diverting our attention from a blue object to a red one at the same actual distance from the eye, and must relax accommodation upon looking back at the blue object."

Red, orange, and yellow usually form a sharp and clear image on the retina—even through distance and haze—while blue and violet tend to appear blurred. According to W. Allen Wallis, yellow is seen as the largest of colors, then white, then red, green, blue, with black smallest of all. A bright image also tends to "spread out" over the retina, just as a drop of water will creep over the fibers of blotting paper. Thus, bright colors appear large and warm colors appear near.

These phenomena have practical application in posters and packages. Bigness is to be accentuated through the use of light, warm colors. Elements in a design meant to stand out with prominence should advisedly be red, orange, and yellow —and, preferably, be set off against greenish, bluish, or purplish backgrounds of low value; for dimension in color increases from coolness to warmth, from darkness to brightness, and from grayness to purity.

Color Contrast

It is a phenomenon of vision that the stimulation of any one color will bring forth an afterimage of its complement. After looking at red, the eye will experience the sensation of green; and after looking at green, it will experience red. Other such pairs are to be found in yellow and blue, orange and turquoise, purple and yellow-green.

A simple demonstration of the afterimage is illustrated

The Afterimage. Stare at the center of the black circle for about thirty seconds. Then transfer your gaze to the small dot.

herewith. The illusion is common enough, although the author frequently encounters persons who have never been aware of it. The afterimage has been explained on a photochemical basis and, more recently, as a phenomenon taking place in the brain. Hypnotized subjects have been able to "see" afterimages merely through the *suggestion* of certain stimuli (no colors really being present). And this has happened even among persons who in a waking state have not known that such a thing as the afterimage existed!

When perfect opposites are utilized, one next the other, the effect is naturally exciting, for the afterimages of the hues will enhance each other. This will hold true when the areas are of relatively large size. If the two colors are at all confused or visually blended, an opposite result may take place, one color canceling the other. The principle involved is separately illustrated in black and white.

Color contrast and diffusion. Colors of strong contrast are startling to the eye when isolated from each other; but they cancel each other when diffused.

When two colors are not exact complements, both may appear slightly modified in aspect. With red and yellow, for example, the green afterimage of the red may give the yellow a slight lemon tinge. Likewise, the blue afterimage of the yellow may make the red purplish. This illusion has its value in displaying colored merchandise, and a few principles and case histories will be quoted in a later chapter.

It is also true that brightness will accentuate darkness, just as darkness will accentuate brightness. White looks whiter on

Brightness-contrast. Both arrows are identical in gray tone. Brightness is accentuated by darkness, and darkness by brightness.

black than it does on gray. Couple this illusion with that of the afterimage and it is easy to appreciate how striking effects may be engineered through careful plan and formula.

Color Constancy

Perhaps the most unusual of all visual phenomena is that termed color constancy. This is the curious and amazing ability of the eye to see the world of color as normal under widely different conditions of illumination. White surfaces, in particular, hold their appearance, whether showered with brilliant light or clouded in dimness, and whether seen in normal white light or in strongly hued chromatic light. As an example, when one is sitting near a window, a piece of white paper held in the hand will be unmistakably white. So, too, will be the face of a clock at the far end of the room. Both surfaces will be white in sensation—although, under the same circumstance, the camera would photograph one as white and one as gray.

For this reason, there is no such thing as underexposure or overexposure in human vision—one of the chief differences between eyes and cameras. Under bright and dim light, however, one phenomenon is evident. Although white will remain white, medium and deep tones of color will tend to blend together and to resemble black; and in very somber darkness (where color constancy may be incomplete), the eye will have a grasp on two major values only: extreme lightness will be perceived as one value, and all other medium and deep tones will melt together to comprise a second value bearing likeness to black.

One of the first observations of color constancy was that of Bouguer, in 1760: "The sensibility of the eye is independent of the intensity of the light." David Katz in his remarkable book, *The World of Colour*, has written exhaustively on the subject. Color constancy has been observed not only in man but in lower animals.

In certain fish, capable of color change, background alone seems to be of importance to the reactions; degree of total illumination is of minor consequence. Walls writes, "The shade assumed by the skin of the fish is always (unless the intensity of the incident light is very low or extremely high) in accordance with the albedo of the substrate—the percentage of incident light which the substrate reflects." Katz tells of similar happenings with hens. When the hens were trained to select only white grains of rice and to reject colored grains, they continued to do so when their environment was flooded with strongly hued light—and when, in reality, the light reflected by the grains was strongly chromatic.

For the purpose of this book, the incompleteness of color constancy under low illumination suggests a vital precaution in package design. A design worked out in well-spaced values that shade from light to medium to dark may not hold this neat appearance on the dimly lit shelf of a store. Meticulous effort on the part of the artist may turn out to be largely a waste of time.

Color constancy under chromatic light also has its practical applications. In a room illuminated by red light, for example, the eye will struggle to hold a normal sense of color value. White surfaces (now reflecting red light) will nonetheless still appear white. Other colors (red, green, blue) will likewise retain hints of their reality. In the glorification of certain merchandise, chiefly by display, chromatic light (tinted, of course, and not too vivid) may be used to pervade show windows and interiors with a pleasant blush of charm—and normal values will suffer but little. Show windows with summer fashions under pale blue light, or with winter fashions under pale orange light, will hold appeal and seem psychologically appropriate to the seasons. Dressing rooms, cocktail lounges, rest rooms may have pinkish light to flatter human complexion. All such devices have excellent possibilities in modern business. For chromatic light of agreeable tint (preferably bluish, purplish, pinkish, or amber) gives an interior or a display that same fascination experienced at sunset, during the lull before a storm, or in the mellowness of Indian summer. Obviously, greenish or cold yellowish lights are undesirable. They turn the flesh gray and endow it with a cadaverous pallor.

In experimenting with color constancy and the action of colored light, Harry Helson noted that (*a*) light surfaces (gray) took on the hue of the light source; (*b*) medium tones remained apparently normal; (*c*) dark surfaces were tinged with a hue complementary to the color of the light source. For the most part, chromatic illumination gives the world a softer and more filmy aspect. It has an atmospheric beauty that most human beings enjoy and relish. Pervading a show window or a display, it may be used for the predominant effect, while some of the merchandise is spotted with clear light to reveal its true appearance.

Color in Practice

So much, then, for the human nature of vision, a subject that could occupy several chapters of its own. Concerning ourselves again with facts and figures, we will set forth practical rules for the use of color in advertising, packaging, and displays.

Of seeing, James Southall has written: "Good and reliable eyesight is a faculty that is acquired only by a long process of training, practice, and experience. Adult vision is the result of an accumulation of observations and associations of ideas of all sorts and is, therefore, quite different from the untutored vision of an infant who has not yet learned to focus and adjust his eyes and to interpret correctly what he sees. Much of our young lives is unconsciously spent in obtaining and coordinating a vast amount of data about our environment, and each of us has to learn to use his eyes to see just as he has to learn to use his legs to walk and his tongue to talk."

One hears copy writers speak of the need for reaching mass audiences with words and phrases geared to the intelligence of young age levels (simple words, mind you, and not necessarily inane and maudlin ideas). This is also good policy for color. Too much sophistication, too much artfulness is often missed by simple emotions, just as fancy sentences may be missed by simple brains. Though the primitive and elemental may seem banal at times, it may, with discretion, hurt no one's feelings and ring clear in its appeal.

I am reminded of an experience in the design of a peanut-butter container. An attractive layout, colored in hues of pale cream, tan, brown, and green was developed after months of artistic deduction. The final selection of this among other similar packages was made through the competitive process of submitting various "dummies" to consumers over the counters of retail stores. The public jury, after leisurely inspection, voted for the combination mentioned above.

Yet a later package, designed on good scientific premises (visibility, legibility, etc.) and without the benefit of consumer opinion, greatly outsold the more prized design. This was red, white, and blue. High on a shelf, with no one asked to say whether or not it was pleasing, it had the necessary lighthouse qualities to compel attention and to set hands reaching for it.

Chapter 9

MORE POWER TO ADVERTISING

ADVERTISING is one of the major industries of America. It is a consistent and successful user of color. Yet, while many practical measurements of the value of color have been attempted, most companies are satisfied to employ it without demanding proof. They know the human appeal of the spectrum and are willing to promote it accordingly. Color sells. Its benefits are to be admitted, whether or not specific data are offered in evidence.

However, there is wisdom in gathering as many facts as possible. After all, color is expensive. Business ought to know what to expect for its investment. In a recent survey by the Color Research Department of the Eagle Printing Ink Company, among 56 users of direct mail, 40 of them endorsed color without having exact figures to present; 9 were rather sure of themselves, having precise data to submit; 5 had no comment; and 2 out of the 56 definitely stated that color had no advantage over black and white. More research ought to be attempted, to encourage an intelligent exploitation of color backed up by a realistic knowledge of its power.

Direct Mail

In the field of direct mail, color tests are rather conveniently made. Many organizations have sought to measure the effectiveness of hue. The following data have been taken from various sources. The author is particularly grateful for the help of George D. Gaw and the Direct Mail Research Institute of Chicago.

On the negative side, one manufacturer reported on three tests of four-color process against black. He states: "In each case where such a test was made we found that the number of orders per 1,000 circulars resulting from the color job was lower than the number resulting from the black and white job."

It is perhaps fair to consider these tests accurate and to let them serve as a caution that color does not always wave a magic wand. Too frequently, the direct-mail advertiser uses color carelessly, if not extravagantly, feeling that for some occult reason it will be extremely profitable to him. Just as good copy is needed to sell merchandise, so careful and deliberate attention should be paid to layout, typography, illustration—and color. And the more research is conducted to measure results, the more scientific will advertising become.

The possibilities are exciting. An overwhelming confidence in the value of color ought to inspire a quest for statistics. Some colors may be better than others. Some applications may be more appropriate in featuring certain merchandise. To study color in this light is to be in a position to audit the effectiveness of advertising and to get the most profit from expenditures.

As to envelopes, a manufacturer of business machines found that blue pulled 7.8 per cent against 3.1 per cent for white—better than two to one. Yellow pulled 6.8 per cent, goldenrod 6.4 per cent, green 6 per cent, and pink 5.8 per cent.

A paper company in testing the relative merits of five different envelope colors (printed with both one-color and two-color corner cards) found pink to be five times as productive as white. Yellow stock was second best, and green stock third.

A publisher in testing the relative effectiveness of color, in both letterheads and envelopes, worked out a series of test pieces featuring the primary appeals of white, red, yellow, green, blue. His research confirmed the fact that a combination of black and red on white (formerly used) was a good one. However, certain of his lists gave preference to a scheme involving dark blue inks on pale blue stock, and he profited in consequence. The complete test was as follows, 100 per cent being arbitrarily set for black on white in order to establish a comparison.

Colors	Per Cent	Colors	Per Cent
Two shades of blue on blue	108	Blue and red on white	89
Black and red on white	105	Purple and green on green	89
Black on white	100	Brown and orange on canary	85
Blue and red on pink	92		

Another publisher found that two colors on a mailing piece increased returns from 10 to 15 per cent over black and white; the extra cost was only 1 per cent. In a similar instance, pink letterheads were 80 per cent better than white. Green was second best, buff third, and blue fourth.

A manufacturer of building materials discovered that a two-color letterhead produced only 1.1 per cent more returns than black and white. Because each inquiry represented a sizable prospect in dollar volume, the extra expense was justified and showed great profit. Another company noted that two-color literature pulled 20 per cent better than black and white and produced orders at a lower unit cost.

In a test of outside envelopes, goldenrod produced 21.42 per cent of orders, pink 17.83 per cent, green 17.82 per cent, white 17.29 per cent, kraft 15.89 per cent. The old envelope previously used (color not stated) produced 9.75 per cent.

Return envelopes were similarly tested by a publisher among some 100,000 prospects. The result was 26.2 per cent for pink, 26.1 per cent for blue, 25.1 per cent for yellow, 22.6 per cent for green. Another company found goldenrod best, pink second, and green third.

A milling company, testing return cards, found cherry red brought in 50.6 per cent of orders, blue 32.7 per cent, and white 16.7 per cent. An insurance company found bright blue ink on white stock best for application cards. Order blanks with orange borders were better than combinations of other colors.

An organization selling an investment service found that a blue reply card pulled 1.64 per cent against 1.57 per cent for buff, 1.55 per cent for rose, 1.51 per cent for green, and 1.50 per cent for a government card. A manufacturer of electrical equipment found a cherry red card to be best, green second, and orange third. So-called c.o.d. cards in color have also been found superior to stamped government post cards.

A manufacturer of inked multigraph ribbons produces 20 per cent of them in color today, as against 1 per cent five years ago.

While such case histories indicate that color offers profitable returns, not enough advertisers attempt to measure it carefully. More tangible proof is essential if direct mail is to progress on a basis of science rather than speculation. Guesses and mere generalizations are not enough.

The Direct-mail Technique

In an article for *Printers' Ink* (Sept. 15, 1938), Frank Egner offers some pertinent notes on the functions and values of color in direct mail. Bright, clear hues always seem to be best. However, restraint is necessary. "Some mailing pieces are definitely ruined in effectiveness by too much color. Pictures of product or package in a circular should, of course, be reproduced as exactly as seems economically feasible. But in general the mail-order man tries to remember that he is using color only for contrast and emphasis and that if he uses too much there will be no contrast and no emphasis. Color for rules, for important headlines, for occasional initial letters, is usually sufficient for his purpose. When color is massed, as for instance in tint blocks, the lighter colors should be used. Color always requires more white space around it than does black ink, because no color, however heavy it may be, offers so complete a contrast to white paper as does black."

Egner has presented seven rules which are practical to observe.

1. Always use a second color in sales letters.
2. Use a colored stock or a second color in order cards.
3. Print dark colors on light backgrounds, rather than light on dark.
4. In general, choose the reds and the combinations of yellow and red and yellow and blue.
5. Use color sparingly for best effect.
6. Arrange for plenty of white space around masses of color.

7. Bear in mind that a colored stock will often have the advantage of a second color with less expense.

Space Advertising

To quote from another survey conducted by the Eagle Color Research Department with the help of the author, here are more facts on the effectiveness of color, this time with reference to space advertising. Perhaps the most comprehensive study has been that of the Association of National Advertisers, "Color in Magazine Advertising," issued in 1931. This 88-page report is both complete and well stocked with figures. Although it is more than ten years old, much of its information is still pertinent.

According to a questionnaire issued when the report was prepared, 27.1 per cent of magazine advertisers used color because of the added values it gave a product; 23.7 per cent endorsed it for its attention powers. The main argument against its use (25 per cent of advertisers) was high cost. Obviously, color is more expensive than black and white. In consequence, its use is not to be justified unless it pays both its cost and a profit beside.

In the ANA report, certain research of Daniel Starch is quoted. In an analysis of 5,000,000 inquiries, involving 2,349 advertisements run by 163 firms, color brought an average of 53 per cent more returns per 100,000 circulation than did black and white. To quote the report: "It is well to remember that the very magnitude of Dr. Starch's study does not make it conclusive evidence of the relative pulling power of color to advertise any particular product. In the first place it covers twelve years and . . . conditions have changed markedly during the last decade or so. Many more advertisers are using color today than ever before, so that an average over twelve years must necessarily cover a period which begins when color competition was limited and extends through a period when it has become intense."

Starch also discovered that color did not pull equally well for all types of products. How this 53 per cent advantage, which apparently existed ten years ago, ranks today will be indicated later.

In mail-order catalogue selling, the ANA report quotes from an article in *Printers' Ink Monthly.*

"Sears, Roebuck & Co., Montgomery Ward and the National Cloak & Suit Co., with experience extending over many years, find that color, in a number of instances, outpulls black and white about six to one."

"In testing out the effectiveness of color, a mail-order company reproduced an article in natural colors for half the run and in black and white for the other half. The color page pulled fifteen times better than the black and white page."

"Another impressive experience was that of a stove company which found that while its color advertising cost 70 per cent more than black and white, the returns were 395 per cent greater."

Advertising Today

This is an age of color. The rapid growth of color processes in reproduction, engraving, and printing and the vast increase of color revenue in magazines and newspapers offer good evidence that color is profitable to the advertiser. In 1907 *The Saturday Evening Post* ran 5 per cent of its advertising pages in color. In 1922, the percentage had gone to 28. Twelve years ago an average issue of the *Post* had 68 per cent colored pages. The period of the depression caused some curtailment, which since has been pretty well restored. Today (1944) approximately 69 per cent ot the full pages are again in color.

As to revenue, the figure for color in 1913 in the *Ladies' Home Journal* was 11 per cent. In 1938 it was 49.6 per cent. In *The Saturday Evening Post* (1938) it was about 52.5 per cent.

In the mail-order catalogue, Sears, Roebuck and Company ran 94 colored pages in 1914 and 246 in 1941. (The figure was less during the war period because of restrictions and the scarcity of consumer goods.)

The Starch Reports

The organization of Daniel Starch has for many years made constant measurements ot the value of color in space advertising, and an early report has been quoted. Today,

Starch offers further facts based on more recent analyses. The conclusion is that "Color materially increases visibility effectiveness of advertisements."

In a study of two identical advertisements, which ran both in black and white and in color, Starch reports the following:

Among both men and women a black-and-white page had an average visibility of 30.7 per cent; a four-color page, 50.5 per cent.

Among men the visibility of the black-and-white page was 30.9 per cent; the four-color page, 46.4 per cent.

Among women the visibility of the black-and-white page was 31.1 per cent; the four-color page, 52.1 per cent.

These percentages were for products in 12 classifications and included advertisements for such things as automobiles, clothing, food, household furnishings, liquor, cosmetics, etc.

Some time ago, E. T. Gundlach made an intelligent analysis of the value of color in space advertising in an article for *Printers' Ink* (May 17, 1940). He stated that, as a mere attention getter, color increased interest from 25 per cent to a normal 33⅓ per cent and over. "In some cases, because of the nature of the merchandise, color has proved its value at 300 per cent, possibly 400 per cent."

Gundlach presented a list of color values based on experience and record, as follows:

Color as an attention getter, when it is not in marked contrast with its surroundings, has a potential value of from 20 to 25 per cent.

When it is in marked contrast with black and white, its value is from 33⅓ to 50 per cent.

On a back page in a magazine, with "glistening colors," its value is from 33⅓ to 50 per cent.

However, if only one member of a complete range of merchandise is illustrated in color, the observer may be disappointed and the value of the color may be zero (or even worse than black and white).

In showing merchandise, if the color in the advertisement is merely incidental (such as the color of a package), the value may run from 0 to 10 per cent.

When the color is integral with the product and really glorifies it, the value may run well over 100 per cent.

Case Histories—Publishers

Here are individual notes from the experience of various companies.

A publisher writes, "We have estimated the value [of color] by setting our four-color rate per page at a premium of 38.4 per cent over black and white."

A trade journal reports that a few years ago colored advertisements were rather few and far between. Today 12 per cent of his national accounts use color.

Another publisher in the business field writes, "We have one advertiser using eight four-color inserts during the year and four black and white ads. We have been told by him that one four-color insert produces as many inquiries as the total from the four black and white ads. We feel that it is a safe assumption to say that color advertising of this type produces 4 to 1 over black and white space in our field."

A publisher in the sporting field quotes from two similar advertisements for books, one in black and white and one in full color. He states, "Our records show that the use of color has brought us approximately 50 per cent greater returns than the black and white page."

Case Histories—Agencies

One prominent agency that has checked the reader observation of men and women reports:

"An average of four magazines, three leading weeklies and one monthly magazine . . . shows a combined average of observation for *men* as follows:"

Colors	Per Cent
One page, one color	36
One page, two colors	40
One page, four colors	51

For *women:*

Colors	Per Cent
One page, one color	34
One page, two colors	28
One page, four colors	52

One failure for color is to be noted in the two-color page read by women. Was this a weakness inherent in color itself or merely in the particular choice made?

Another large agency points out that the value of color may vary in different magazines. From an analysis of four weekly publications, here is the percentage of increase in *visibility* of four-color advertisements over black and white.

Magazines	Per Cent
Magazine *A*	88
Magazine *B*	18
Magazine *C*	14
Magazine *D*	1

The agency writes, "If you believe these figures for *observation*, there is good reason for using color in magazines *A*, *B*, and *C*, but not in magazine *D*."

"Friends of color serve it best when they use it judiciously according to specific needs. Color has to be mixed, as a famous painter once remarked, with brains."

A New York agency writes, "We recently conducted a test of four-colors vs. black and white to find, in this particular case at least, that color produced 44 per cent more returns per thousand circulation than black and white, but the advertisement was one that lent itself particularly well to color treatment."

A Chicago agency writes, "We find that two-color advertisements have a greater attention-value than the same advertisements in black and white, but that the black and white advertisements receive greater reading time. A layout that includes a four-color illustration has a greater attention-value than has the same layout in black and white. A four-color advertisement also has greater readership value than the same advertisement in black and white."

Case Histories—Advertisers

A national advertiser, offering a give-away picture, states that full-color reproductions drew four times as many inquiries as did black and white.

Another manufacturer in the clothing field writes, "We judge our color advertisements by two methods. One by the consumer inquiries that we receive and the other is of course on the direct sales which can be attributed to advertising. An example of the consumer inquiry method is as follows: One black and white advertisement in *The Saturday Evening Post* pulled 338 inquiries, and the four-color advertisement in the same magazine pulled 1,334. The advertisements were different sizes, which means that the four-color ad had an additional advantage here. But as a half-page ad is supposed to draw the best ratio of inquiries to cost of space we feel that the advantage of the four-color full page ad was not as tremendous as it might seem.

"As for direct advertising results, we know that sales figures have increased as we have used color in our advertising, because it gives us an opportunity to merchandise our advertising, show color, and pull promotions for store buyers who are prone to stock more merchandise to take care of the consumer demand which will result from the promotions.

"In our magazine advertisements we always try to use two colors, red and blue. Our particular merchandise is colorful and for men we use blue and for girls we use red. This was somewhat changed this year to show the new color styles, but we find the best results from the use of the colors mentioned."

A maker of farm machinery states, "We have had tangible proof of the value of color. Our inquiries are about three times greater than with black and white advertisements. Today we use color wherever we can get it, even in our advertisements to dealers, and our cost per inquiry has gone down, and since going to color we have spent a smaller percentage of our sales for advertising than we had spent formerly."

A food packer states that he receives more *readership* per dollar invested with black-and-white and two-color advertisements than he does with four colors. However, "the merchandising value of the four-color pages was sufficiently far above that of the two-color or one-color pages to make them our choice in the advertising of food products."

A textile manufacturer quotes an interesting case history. "Years ago we used to offer samples of wash fabrics, sometimes free, and sometimes for a stamp. We also sold a small novelty featuring another of our products. On the samples I could count on a minimum of 10,000 inquiries per color page, while black and white space using full page or fractional was always uncertain and always a higher cost per inquiry. On the novelty, which sold for 20 cents and sometimes 25 cents and which never was highlighted in the headlines or. text, but limited to a small mention in the lower corner of the ad, I could almost tell in advance the number of orders we would get when we used color, but we never could predict with any confidence when we used black and white."

Color Technique

Two facts seem apparent: (1) that color increases the effectiveness of advertising, but (2) that its value is largely dependent on a wise and appropriate application of its powers. Merely to use color for the sake of color is, of course, not enough.

Thus, to get the most out of the spectrum, the following suggestions may be considered. They may serve to inspire reason and purpose in advertising design and to put color to work in a sensible way.

1. Attention-value: Color will catch the eye where mere black and white may fail. However, because color is compelling, it may also be distracting. Some tests have shown that, whereas color may have higher noticeability, black and white have greater readership. This means that care must be taken in the typographical layout. Color should be simple and appropriate, and should be tied in to hold the advertisement together. It should invite readership rather than discourage it.

2. Realism: In many products, color is important in the display of a product. As a dimension, it may be as descriptive as size, weight, price, etc. Here, perhaps, is the most obvious use of color—and one of the best—to glorify a product in its full reality.

3. Identity: Colors are more easily retained in the memory than are words or symbols. Used to identify a product or a service, and persistently employed to build up recognition, color has potent force, and primitive hues are generally superior to intermediate or modified colors.

4. Psychology: Because of the many curious mental and emotional associations of color, subtle applications are possi-

ble. The use of color to imply warmth, coolness, cleanliness, etc., may be extremely effective and may well support copy efforts.

5. BEAUTY: Color has intrinsic appeal. Glorified in drawing or photograph, it will command interest, whether or not it has a direct bearing on the copy message. This utility, distinct from attention-value, demands greater artistic skill and has produced some of the best advertisements ever composed —as aesthetically satisfying as any good work of art.

Chapter 10

PACKAGES, DISPLAYS, INTERIORS

NOW we proceed to the business of packages, displays, and interiors, and to more facts and case histories. Much science and theory have been considered before, in the chapter on vision. Let this be supported with reports from the field and with principles that draw from the fertile ground of fact.

As is true in the field of space advertising, most business executives recognize the importance of color in packaging. But here, again, it is unfortunate that few well-organized tests have been made to prove the merits of any particular design or color scheme over another, although in recent years packaging has become a high art. New ideas, new devices have been put to work by almost every organization. The growth of packaged merchandise, the supermarket, have demonstrated that successful marketing demands not only a quality product at a fair price, but real eye appeal and attention-value. Design and color, rightly and strategically handled, help to sell merchandise and support well the attendant efforts put into advertising, posters, displays, dealer helps, and the like.

While, perhaps, it is not possible to write ideal specifications for the use of color in packaging—because of the widely different requirements of different products—modern business may benefit from the recitation of several case histories and experiences. It may then be possible to reach more intelligent conclusions as to any individual packaging problem. Many of the following data have been taken from a special survey conducted, with the cooperation of the author, by the Color Research Department of the Eagle Printing Ink Company.

Case Records

To discuss foods first, one Western maker of food products (mayonnaise, salad dressing, etc.) gives a very intelligent report. He states that records and charts have been kept to measure the sales increases brought about with new designs and color combinations. He writes, "We have found that dark blue and bright red are the outstanding colors. Red, although a greater impulse color, has moved into second place in our operation." To capitalize the apparent value of these two hues, he recently produced labels incorporating both colors on white stock. "We have been more successful than we ever imagined; the response was immediate."

The combination of red and blue has also been found most effective by a group of grocery-store chains using private labels.

Red and white (with gold and black) have, by one national canner, been ranked eminently high for both attention and memory value.

A baker reports, "The color ranges that we find most satisfactory are the warm colors, such as orange, red, brown, and light blue. We have no comparative figures by volume, but the greatest amount of our goods are sold in the orange or red-orange shades."

As related to foods, several interesting facts have been assembled by Raymond Franzen and Thomas Meloy and published in the January, 1941, issue of *Modern Packaging.*

In an analysis of 1,755 labels of 120 leading canners, 49.6 per cent were found to use 4 colors in labels; 37.2 per cent used 5 colors; 5.1 per cent used 6 colors; 4.7 per cent used 3 colors; 2.2 per cent used 2 colors; and 1.1 per cent used only one impression.

In the one-color group, blue ranked highest, brown next.

Red and blue were most popular in two-color labels, red and black second, red and green third, and yellow and blue fourth.

The leading combinations in three-color labels were red, yellow, blue; red, blue, gold; red, green, gold; red, yellow, black.

In four-color and five-color labels, red, yellow, blue, black, combined with light blue constituted the choice of the vast majority.

Because many food labels show realistic illustrations, the process colors (red-yellow-blue) are naturally in predominance. However, if there is any strong predilection beyond the primaries, it is overwhelmingly in favor of light blue "as a supplementary color to provide special emphasis for borders, lettering, letter shading, or some other detail of the label."

Incidentally, blue-lined containers have been found most effective in selling white eggs, while white-lined containers sell the most brown eggs.

One of America's leading soap manufacturers quotes his experience in testing consumer reactions to color. He writes, "We have run extensive consumer interviews to determine the preference of housewives for colors in packages, but have never tried to extend these reactions beyond the particular items tested. When our present soap wrapper was adopted several years ago, we had made many interviews and had conclusive evidence of a preference for green on a toilet soap wrapper. There was not much indication of design preference as long as the basic color was green."

A maker of paste floor wax states that an increase of approximately 25 per cent in sales followed the introduction of a new design and color effect. The colors used were red, yellow, and black.

A manufacturer of razor blades writes, "When we originally went into the blade business about five and a half years ago, our package was a drab brown. From that package we went to one which was a green and blue color combination, and finally, about three years ago, we evolved our present package, a blue and red. . . . . We definitely know that the present color of our package has given us greater display value and has influenced the dealer to place the package on his counter."

The Ideal Package

It is apparent that research pays and that it will help to eliminate guesswork in the use of color in packaging. Yet, as has been cautioned before, a right and scientific approach may be a better guide to design efforts than is too much leisurely consultation with consumers. The personal viewpoint here is far less vital than a resourceful exploitation of those phenomena encountered in a study of vision.

Red and blue, one a color of high recognition, the other a color of universal appeal, are predominant and rank almost equal in preference.

Yellow, the color of highest visibility in the spectrum, naturally finds widespread use.

Green, which also ranks high in preference, seems to lend itself chiefly to certain restricted uses.

Beyond these four colors (recognized as primary by the psychologist), few other hues or tones have been employed in the majority of successful packages. Probably there is a good reason for this. While odd shades, blue-greens, yellow-greens, lavender, pink, etc., may be individually and intrinsically beautiful, they lack primitive and primary qualities and hence fail either to compel the eye or to impress themselves on the memory.

Finally, here is a practical check list of objectives in the design and coloring of a package. The six points, if kept in mind in the order given, should aid in achieving an effective and profitable result.

1. The first duty of the package is to command the eye. Here color is perhaps most vital.

2. After it has caught the eye, the second duty of the package is to tell what it is and to establish its identity. Here typography and design are important.

3. The third duty is for the design and the color scheme to be appropriate to the product contained. What factors in design and color seem to lend themselves best to a particular item?

4. The fourth duty of the package is to please the eye and the emotions. A startling package that catches the eye but fails later to please is not so good a package as is one that achieves both.

5. The fifth duty of the package is to invite handling and further examination. This is where neat details count and where the customer is urged to open his or her pocketbook.

6. The sixth duty of the package is to be well constructed, durable, and handy in use.

Displays

When the display itself is meant to be the object of attention, as in the case of a counter card or a window setup, the problem of a color effect corresponds with that encountered in packaging. Simplicity, high visibility, and attention-value —these are the requirements.

Where, however, the display concerns the featuring of merchandise, different strategies may be required. In the concentration of visual interest there are a number of devices to employ.

Brightness against darkness.

Warm color against cool color.

Pure color against gray color.

Detail and texture against filminess.

Form against (or surrounded by) plain space.

All these contrasts may be applied separately or together. Further interest may then be stimulated through careful choice of hue, keeping the primary color for the target area and letting the background blend out into subtle and softer tones, which lend atmosphere and yet introduce no major distraction.

With illumination, dramatic effects may be heightened. Brightness contrasts may be accentuated with focused spots of intense light. Shadows, rounded forms, space relationships may be enhanced through the use of chromatic lights directed from different angles. Lights, of course, differ in mixture from similar combinations of pigments—red and green forming yellow; green and violet forming turquoise blue; and red and violet forming magenta. Thus, an object standing before a background will, if illuminated by red on one side and green on the other, appear yellowish. Shadows, however, will be deep red (where the green light is cut off) and deep green (where the red light is cut off). By carefully spacing red, green, and violet lights, a mixture of white may be effected, to give the displayed object an almost normal appearance; yet, all the shadows and high lights may sparkle with iridescent hues having endless variety.

Interiors

The use of color in interiors is a big subject. In merchandising, the technique usually involves the application of hue as an accompaniment to the things being sold In this regard it cannot be too dramatic in and of itself, lest it defeat its purpose. However, unusual and remarkable effects are possible. The following chapter will describe the author's experience (chiefly during the war period) in the scientific application of color to industry. Here, in studying problems of production, eyestrain, safety, morale, many useful lessons have been learned that have much to do with mass reactions and therefore with control and influence over human moods.

For the present, however, although we are less concerned with industrial efficiency than with salesmanship and merchandising, a few general points will be in order as a prelude to the art of interior decoration—aesthetic, commercial, or industrial.

Seeing is dependent upon two factors, illumination and color. The first is the source of stimulation, the second is the actual thing perceived. However, the specification of color by artists, designers, decorators, or lighting engineers is not too well understood either in theory or in practice; for, while the color (and therefore the merchandise, window, store interior, etc.) seen by the eye is dependent upon light, the two are curiously independent.

Ewald Hering has written, "Seeing is not a matter of looking at light waves as such, but of looking at external things mediated by these waves." The eye rarely sees light itself; it sees the objects and areas that light reveals. Nonetheless, human impressions of illumination are gained primarily from the appearance of surfaces within the field of view. One does not have to glance at a light bulb to know that a room is pervaded by illumination.

These remarks may seem academic, but they lead to a viewpoint that regards color not alone as a thing of beauty in itself but also as a potent factor in influencing judgments of illumination, for interior decoration usually involves large areas that enclose the entire vision of the observer. The designer has a double force at his command—the interest of color in particular and of illumination in general.

The point may be clarified in talking about brightness. The reaction of an individual to a light environment is one that draws attention to the room in general. The reaction to a dim environment is one that draws attention to details, to things within the room. Strong and dim *illumination* (colors the same) will cause these two differences in attitude. And more significant, light and dark *colors* (illumination the same) will do likewise.

There is an explanation for this psychological phenomenon, and a clue to effective merchandising. To begin with, the eye is always attracted to brightness, as against darkness. When products are exposed before a light wall, the wall may seriously compete for attention. Yet, by keeping illumination at the same level and lowering the lightness of the wall with a deeper color, the attention may be shifted to the product (where business naturally wants it to be concentrated).

Store decoration, therefore, has these ideal objectives:

1. Strong and brilliant lighting, particularly around the entrance.
2. A sense of brightness as a first impression upon entering the store. This perhaps means white or pale ceilings and upper walls, to encourage the eye to make a sweeping examination of the whole interior.
3. Brilliant areas or details at far ends of aisles—as color "magnets," to break up monotony and invite greater store traffic and circulation.
4. Deeper tones behind merchandise, properly balanced in hue and value to afford high visibility to the products on display.

For general color treatment let us offer one tried innovation. This involves the device of using color to simulate light and of endowing an interior with a mild and mellow atmosphere bordering on illusion. Where large areas are involved (upper walls and ceilings), pale grayish tints with a glint of hue will effect strange results. Because such delicate colors will lack an obvious hue, the eye may miss the fact that it is seeing paint and may unconsciously gain the impression of chromatic illumination. Cool light will be suggested where the wall tones are greenish or bluish, and warm light where they are yellowish or pinkish. Yet, all merchandise will itself appear entirely normal, being showered with light that is not seriously distorted. This effect is made possible by the fact that the larger areas within the field of view are tinted, the eye therefore assuming that the light itself is tinted.

Further, by using a variety of these pale grayish tints in different sections of the store or on different floors, one may treat the customer to a pleasant change of illumination quality, now pinkish, now bluish, now yellowish, now greenish. The emotional effect is agreeable indeed and, because the colors are not aggressive, they offer no competition with merchandise. Rather, they create a sort of illusory atmosphere that affects the mood without in the least confusing the eye.

To Feature Merchandise

To demonstrate how intelligent specifications for color may be written, let us now describe the theoretical and practical development of a dress shop. This will afford an opportunity to explain how color in its relation to merchandising problems may be analyzed and solved in reasonable and orderly fashion.

What is to be accomplished? The customer is to look her best. Her complexion and her figure are to be flattered. Every dress, no matter what may be its color or style, is to reveal her best features when she tries it on. She is to be satisfied as quickly as possible, so that a prompt sale will be effected.

To execute this assignment, the first question to be settled is that of an appropriate key hue. Through the phenomenon of the afterimage (previously described) it is possible to locate a color which, under the given illumination of the store, will complement what might be termed an ideal flesh tone. Normal flesh is a somewhat muddy affair, pinkish to a slight degree but mainly grayish, tannish, and greenish (that is why the ladies use cosmetics). The ideal tone, seen on the magazine cover or in that famous "high-school" complexion, is obviously pinkish or pinkish tan. A youthful blush of the sort finds its complement in blue-green (on the turquoise side). The afterimage of this hue causes average flesh to pick up in luminosity and to appear more youthful.

So blue-green is settled upon to flatter the skin and tinge it with a bloom of pink. Now, to secure the right tone, the brightness of the blue-green is balanced to reach a level of value on which black and white will be equally conspicuous. Next, this value is again shifted toward gray, to reach a tone slightly blue-green, slightly gray, and halfway in lightness between black and white. The result strikes an average among most colors. Upon looking in the mirror and being silhouetted against this special tone, the customer sees (*a*) her complexion enhanced; (*b*) her dress, whether it be light or dark, pure or gray, red, yellow, green, or blue, well defined upon her figure.

Moreover, in her dressing room the one wall opposite the mirror is of the same special blue-green. The side walls, however, within arm's reach and active in the reflection of light, are tinted coral; and the lighting fixture is so designed as to cast some of this coral hue directly over her face.

Store decoration is to be approached from many angles, and the functional method just described is but one road. I recommend it, however, for its basic purpose, reserving for the individual designer the right to add femininity and frills, to feature period styles, textures, and materials, as he sees fit. After all, if he is going to use color, he might just as well have certain objectives in view.

Practically every item of merchandise is to be analyzed and displayed through like procedures. The carpet in a shoe department should probably not have a design upon it; and, because most of the shoes sold are black, brown, or white, a medium grayish blue may be the best color.

Glassware and silver may be complemented not only with right hues but also with soft and dull textures. Suede gloves and felt hats lend themselves to shiny or metallic backgrounds, to assure an agreeable contrast. Purpose is to be built into almost every use of color—and sales profits.

In this *functional* use of color to sell merchandise and to influence human moods, there are other principles to be listed. From an extensive experience in the decoration of schools, hospitals, restaurants, the author has arrived at a number of practical conclusions, which are here passed on.

Of all colors in the spectrum, two seem to be most versatile and livable. Although these hues are not generally chosen on impulse by the consumer; yet, for all-round utility in the decoration of places where people congregate, they seem to have universal qualities that are well adapted to public places, restaurants, rest rooms, theater and hotel lobbies. The first of these is a soft blue-green, the second a bright luminous red-orange or peach color. Psychological research makes clear that the spectrum has two major regions that are endowed with two distinct appeals—the warm and the cool. Red and blue are the typical expressions. These two colors (both pleasing) are opposite in physical action and in emotional meaning. One is the aggressive, importunate hue associated with things positive. The other is cool, passive, and retiring.

Blue-green and peach are subtle forms of the more primitive red and blue. As they are less elemental, their stimulation is not quite so impulsive. They are more protean in aspect, having different tints when seen under natural and under artificial light. This saves them from monotony. Without being obtrusive, they satisfy most taste, representing in turn the refreshing quietude of nature and the proverbial glow of a fireplace—delights that seem to be part and parcel with the human spirit. The author trusts that his conclusion is not too personal with himself. These are not colors that most people will choose when you ask them. Yet, the deduction has a certain amount of scientific foundation. Here we are hitting averages, formulating hues that sidestep individual prejudices and preferences by glorifying the two extreme regions of the spectrum and striking harmonious chords, rather than shrill notes. We are not selling the colors, but employing them to our own ends and purposes.

Niceties of the sort may not be wise policy where merchandise is designed or packages are created, but in interior decoration it is often within the province of management to express its own will.

A restaurant, for example, has the problem of selling food, and food is best eaten where appetites are stout. What can be better than to play upon appetite with appetite colors? One chain of cafeterias found that orange walls back of serving counters inspired higher sales than any other hue. Another increased the sale of salads 25 per cent by using green plates.

The most "edible" colors are a warm red, orange, a pale yellow, peach, a light green, tan, and brown. The functional approach to color in a restaurant, therefore, might automatically dictate a color effect based on these appeals. At least, a better plan of approach would be hard to conceive.

Hospital Decoration

Finally, the technique of color conditioning (as apart from the usual aesthetic conventions of decoration) finds a good example in the manner in which hospitals use color.

1. In the lobby and reception room (seen more by visitors than by patients) strong emotional reactions are defied through the introduction of variety. By the contrast of warm colors against cool ones, light against dark, any precise reaction is discouraged. The eye is treated to a constant shift of scene (thereby keeping the emotions from seizing upon any one mood).

2. In major and minor operating rooms blue-green is standard for walls. The color reduces glare, offers visual relaxation to the surgeon, and holds up the acuity of his eye by directly complementing the red color of blood and tissue.

3. Most service departments are white, to encourage good housekeeping and cleanliness.

4. In private rooms, particularly in the maternity division, warm tones of ivory and flesh (on the aggressive side of the spectrum) stimulate an optimism and cheerfulness favorable to recovery.

5. In rooms and wards for chronic patients, however, blues and greens (on the passive side of the spectrum) offer a more relaxing and retiring environment and are thus kind to those who must be confined for a longer period.

Industrial Design

Industrial design is a vital and lively business, in these days. Through so-called functionalism and streamlining, art and industry are endeavoring to make products more useful and comely, the matter of appearance following that of purpose.

Color is a thing to be carefully engineered. In products such as typewriters, adding machines, instrument panels, machine-shop equipment for factories, washing machines and other commodities for the home, good industrial design suggests an appropriate attention to such factors as visibility and eyestrain.

Unfortunately, this attention is sometimes neglected. Let us quote a few instances. A sewing machine is redesigned and equipped with a small localized light source. Instead of aiding the user, this light produces objectionable eyestrain, trying contrasts in light and dark, and specular reflections that are annoying and fatiguing, if not actually inimical.

Another type of machine (completely changed in aspect from the original form) requires the user to observe his task against empty space. This leaves the whole efficiency of the user—and therefore the efficiency of the machine itself—entirely at the mercy of the environment in which the operation is to be performed. An ideal seeing background should, in fact, have been designed as an inherent part of the device itself.

Many home products, radios, thermometers, clocks, and the like, are often difficult to use because of low visibility. For some strange reason, in their designing only mechanical and aesthetic features seem to have been regarded. Visibility, legibility, light and heat reflection—all matters that lend themselves to direct scientific method—are often overlooked or confused by purely empirical judgments.

A great part of functional design, in truth, is not functional at all, but merely represents a few mental gymnastics on the part of someone who has fairly good deductive faculties. Reasons are too frequently childish. Green, for example, may be used in a certain application because it is supposed to be easy on the eyes, whereas there may be no scientific basis for the conclusion. Red, another hue often endowed with false values, may be used to mark controls or to be conspicuous as a tag, mark, or symbol on an oil burner; yet red, among all colors of the spectrum, will be the first to fade out in the dim light of a basement.

Aristotle once wrote that a fly had eight legs, and for many centuries the savants of other generations kept repeating the statement.

To set the record as straight as possible, and to be of help to the industrial designer, we offer the next chapter. This is a review of practical experience with color and people in industry, with principles that contribute to human efficiency and comfort, with techniques developed by ophthalmologists and authorities on seeing. There is much to learn here—and far more still to be learned—that should add to anyone's competence in handling color and people.

Chapter 11

THE ART OF COLOR CONDITIONING

EVERYTHING seen by the human eye is colored. In the home, in factories, stores, offices, schools, environment is important. While this book is essentially concerned with problems of merchandising and advertising, the present chapter will furnish a slight digression, for discussing the relative importance of color as it has to do with physical well-being and efficiency.

The curious power of color to influence human moods is well recognized; yet, because it is a thing emotional, most authority in its use has been left with those who have an aesthetic attitude. Color, however, is by no means a matter that concerns beauty alone. Today it is being *engineered*, in the true sense of the word, to do an effective job in improving the health, morale, and security of the individual.

Those of us who spend a considerable part of a lifetime indoors may or may not recognize the importance of light and color in our surroundings. In nature, our attitude is usually casual. Nature is full of delight, variety, and color. We get along all right with her. However, most of our work —the most critical use of our eyes, brain, and hands—is performed in more artificial surroundings. Having these surroundings right or wrong can make for good or bad results in us.

There are many curious interrelations between ease of seeing and the functioning of the human body. Where there is eyestrain, physical reactions are to be noted in a number of responses, both in the eyes and throughout our systems. Abuse of vision may have tragic consequences.

First of all, we shall list the major causes of eyestrain.

Inadequate or insufficient light.

Low visibility due to wrong color contrasts.

Glare.

Distractions caused by excessive brightness and motion on the outer boundaries of vision.

Severe adaptive changes in the pupil of the eye from light to dark.

Continual changes in accommodation from near to far.

Prolonged convergence of the eyes on near objects or on fine detail.

Lack of convenient and agreeable areas for visual relaxation.

Understand, however, that we are speaking of men and women in the process of work or concentration—as they pack the factories and offices of the land and depend for a livelihood on the expert use of their eyes.

When too many of the above bad conditions exist, the results of eyestrain will be apparent in such manifestations as these:

The pupil of the eye will show severe dilation after periods of work.

The lids of the eyes will blink more rapidly.

There may be a collapse in the "visual-form fields," the outer nerves of the retina growing less sensitive to brightness and motion.

There may be specific fatigue of all muscles controlling the eye.

General fatigue and increased nervous and muscular tension throughout the body may be felt.

Headache and nausea may become habitual.

Psychological irritability may develop, or a bad case of jitters.

It is even likely that worse things may happen. M. Luckiesh in his *Science of Seeing* (incidentally, the very best of books on the subject treated in this chapter) writes: "A recent study of occupational morbidity and mortality by the United States Public Health Service revealed that in one company, with approximately 6,000 workers doing precision assembly of fine parts, almost 80 per cent of the mortality cases in five years involved heart trouble. The trend with the rest of the 59,000 industrial workers whose occupational and illness records were studied was quite similar. It is conceivable that the reflex effects of critical seeing and the prevalence of mortality cases from heart trouble in occupations demanding critical seeing may be related. Certainly, heart failure is a common cause of disability and death. Its cause or causes must also be commonplace. The subnormal conditions under which unnaturally critical and prolonged seeing is performed over the course of years and even lifetimes are worthy of the most serious examination, particularly in the light of the present scourge of heart ailments and failure."

It is also recognized that a wretched seeing condition may be an indirect cause of cataract. Dr. Neville Schuler states that where the eyes are strained to see, a bad train of happenings may occur—"alteration in the secretory part of the ciliary body, defective and deficient secretion, nutritional troubles and finally cataract."

Our attitude may seem to be one of undue anxiety, but it rests upon a desire to dramatize the important role that light and color may play in life.

Color Engineering

The successful engineering of color offers a host of benefits. In a factory, for example, with improved visual efficiency go also these attendant advantages:

Increased production.

Better manual skill.

Lower accident hazards and insurance rates.

Higher standards of plant housekeeping and machine maintenance.

Improved labor morale.

When more suitable working environments are provided, management does a finer job of public relations and achieves a greater respect, not only within the organization but within society and the community itself.

In the engineering of color there are two viewpoints to be observed, and each requires a different approach.

1. In critical seeing, where a worker is expected to concentrate for long periods in one restricted field of view, the best principle is one of creating soft and uniform degrees of brightness.

2. In casual seeing, where the worker is expected to be alert to hazards, to pay attention to traffic zones, to watch out for the unexpected in machines, equipment, and surroundings, a more vivid and contrasting use of color is indicated.

Critical Seeing

As to critical seeing, H. L. Logan writes, "General fatigue from any cause, either physical or psychological, predisposes the subject to discomfort under marked brightness contrasts." A similar conclusion has been reached by Parry Moon: "Investigations of the most diverse kind show that a human being sees best and visual fatigue is reduced to a minimum when the entire field of view is approximately the same luminosity as that to which the fovea [center of the eye] is adapted."

There is very good reason for the conclusions of the above authorities. Where the eye is forced to adjust itself constantly to different brightnesses, it is severely taxed; then vision, production, and safety are made to suffer.

Further, the action of vision is quick from dark to light, and slow from light to dark. Luckiesh states, "In general, it may be said that the pupil contracts to a given degree in about as many seconds as it takes minutes to dilate between the same limits in size." On a machine operation, for example, an adjacent white wall may tend to reduce the size of the pupil opening and hence lower visibility of a dark piece of metal. In this case, the wall will have the major advantage. The eye will be quick in adjusting itself to a bright wall (which is meaningless) and slow in trying to discern scribe marks on a piece of dark steel (which is vital). Where the wall is toned to a slightly darker color, however, major distractions may be eliminated and a good seeing condition will be assured.

Thus, for average working environments involving machines or inspection operations, these features may be planned.

1. The ceiling, out of range of vision, may be white, to assure an ample and efficient flood of illumination from above.

2. Upper walls, above 7 or 8 feet, may also be white, to get the most out of existing light sources.

3. Lower walls and columns may be somewhat darker, with a reflectance factor of about 50 per cent (corresponding to light gray).

4. Where it is practical, floors should be reasonably light in color, with a reflectance factor of 25 per cent or better (corresponding to a medium gray). Light floors, through multiple reflections from above, will add more efficiency to lighting than any other single factor. Walls are seldom important in this regard; but light floors will get the most out of light bulbs and tubes, reduce shadows, increase brightness at low levels, and lessen the hazards of falls.

5. Equipment, tables, machines, etc., should also have a reflectance factor of about 25 per cent or more, depending on the materials (bright for light materials and deeper for dark ones).

Thus, the ideal working condition is established where the whole environment is relatively uniform in brightness. There is absence of glare, of extreme contrasts in light and dark, of deep shadows. Luckiesh writes, "It may be concluded that brightnesses somewhat lower than those of the central field are generally most desirable. All experimental evidence indicates that peripheral brightnesses higher than those of the central field are definitely undesirable." It is of major importance to make sure of sufficient light, to keep ceiling and upper walls white, to hold lower walls a trifle darker, to paint machines in fairly light tones, to make floors reflect as much illumination as is practical for easy maintenance.

Hue Contrast

If the field of view for critical seeing is to be uniform in brightness, then the clear discernment of objects, products, and materials must be accomplished by some other device than by extreme light against dark or dark against light. Good visibility is to be aided by taking advantage of the afterimage.

In metalwork, for example, color contrast becomes extremely useful in emphasizing certain materials. Such contrast is achieved better through color than through severe differences in light and dark. Tan colors with a reflectance factor of about 25 per cent draw forth a strong reaction to the normally bluish cast of steel. Bluish green hues heighten sensitivity for the orange cast of brass and copper. Where a variety of materials is used, medium grays are indicated, preferably those with a tannish or a bluish hue.

Colors on machinery not only offer the advantage of subtle contrast with materials, but they help to individualize unit operations, setting them apart in a crowded interior and placing emphasis at critical points. The same general principles that have been previously described for the merchandising and display of products are here given a similar, though less glamorous, twist.

Casual Seeing

In contrast with the above principles (in which ease of seeing demands a soft and uniform brightness), the application of color to casual seeing needs impulsiveness, intensity, and harsh contrast. Safe practice in plant traffic, for example, demands that trucks, dead ends, pillars, be highly visible; that aisles, stairways, corridors be not only clearly marked but compelling to the eye. Fire-protection equipment, hazardous parts, dangerous zones, traveling conveyors, and the like should be conspicuous and startling enough to arrest indifferent notice. Here, steady concentration of the eyes is not expected; hence, the marking may be vigorous.

A standard safety code for industry has been developed by the Du Pont Company, with the collaboration of the author. Its purpose is clear enough. Safety practice today in the use of color is too frequently pointless and haphazard. A hue like red, or yellow, for example, may be promiscuously applied here and there to mark aisles, fire-protection equipment, machine guards, low beams, and what not. Many of these identifications may be mere "wolf cries" and, as such, may have little meaning to the worker.

In the Du Pont Code, six major standards have been set up. Each color classification has its distinguishing symbol, which may or may not be used in connection with it.

Yellow (black and yellow stripes) is used to mark strike-against, falling, and stumbling hazards. It is to be used on low beams, dead ends, the edges of platforms and pits, crane hooks and crane beams, and all such situations where danger is apparent.

Orange (arrow or triangle) is to be used to mark hazards likely to cut, crush, burn, shock, or otherwise cause severe injury. It should be painted along cutting edges; near rollers, gears, punching, shaping, or forming parts; on the inside of switch boxes (being conspicuous when the door is open). It is also recommended as a finish for the *inside* areas of guards —to "scream" at the eye when such guards are not properly in place.

Green (symbol of cross) is to be used to mark first-aid equipment of all sorts, stretchers, containers for gas masks, etc. It should also be used as a mark on the wall to enable the worker to locate such devices from a distance.

Red (symbol of square) is to be reserved exclusively for fire-protection equipment. It should be painted as an area on the wall behind extinguishers, and on the floor to prevent obstruction.

Blue (symbol of circle) is the precaution hue. In the railroad industry, blue signs and lanterns are used to caution against the moving of cars that are being repaired, emptied, or filled. In a factory, the blue symbol is to be employed to mark any equipment shut down for repair—machines, elevators, ovens, dryers, tanks, boilers, etc. By factory regulation, no one is permitted to operate such equipment without first checking with the person who has put the blue signal in place.

White (gray or black—symbol of star) is the traffic and housekeeping standard. White on dark floors, or black on light floors, is to be used for aisle marks or for storage areas. This white or black mark may be changed to yellow where it runs adjacent to a hazard. Gray is recommended for rubbish and waste receptacles, empty-bottle containers, etc. The achromatic colors, being conspicuous enough but lacking in high attention-value, will thus serve a useful purpose without projecting themselves too prominently within an interior.

The code has been successfully applied in many industries. By means of it, real sense and order are established. The colors seen have meaning and purpose, and the worker may quickly memorize the location of all hazards and safety devices throughout a plant, thus being ready for any emergency.

Psychological Considerations

In the functional use of color in industry, level of brightness and degree of contrast are frankly more important than hue quality itself. In other words, an ideal seeing condition (involving soft and uniform values) may be greenish, bluish, yellowish, or pinkish, as is desired or as special circumstances suggest. The pupil of the eye regulates itself to light intensity, and such regulation will be approximately the same for light green as it will be for light buff.

Color as color, however, is important in building up acuity through complementation for purposes of high visibility, and in establishing definite emotional and psychological effects.

On this latter point, color offers many agreeable plus factors that have intimate bearing on human attitude, mood and morale. Because it appeals to all human beings, it is an effective medium for revitalizing the industrial plant and making the worker safer, happier, and more compatible with his environment.

Here, then, is the so-called psychology of color to be observed in industry.

1. For working areas, colors for the most part should be light in tone and weak in purity, too much richness being a source of distraction, because of strong emotional interest.

2. For critical seeing, brightness is best concentrated and confined to important fields of view on machines, work tables, etc., where the worker is expected to fix his attention. For casual seeing, brilliant color and strong contrast should mark the object or the area that is to be clearly distinguished.

3. Because many industrial tasks require prolonged and trying convergence of the eyes, end walls faced by the employee may be pleasantly colored in soft, grayish blues or greens, to provide desirable areas for visual relaxation.

4. The most suitable colors for walls or machines are soft, grayish greens, blue-greens, blues, tans, buffs. These should be subtle rather than too aggressive. Impulsive colors, while cheerful at first sight, grow extremely monotonous in time.

Also, strong colors tend to weary the eye and to form disturbing afterimages.

5. Where the working environment involves exposures to high temperature, pale green and blue-green are by far the best colors to use. They will afford psychological relief and, by illusion, will tend to lower the apparent temperature of the interior. Where the environment is cold or vaulty, buff, ivory, and yellow tones are practical. They will contribute a warm and sunny atmosphere and apparently increase temperature.

6. For employee facilities, however—washrooms, rest rooms, cafeterias, etc.—stronger color is appropriate and satisfactory. Here the color effect should be in contrast with the rest of the plant. Such a change of pace will relax the mood and help to offset fatigue. Because of the inherent color preferences of men and women, blue is the best color for male facilities and pink or rose for female. Peach, incidentally, has proved to be the most satisfactory color for cafeterias, being considered the most "appetizing" of all light tints.

Case Histories

The technique of applying color in industry may be demonstrated in a practical way by referring to a few typical case histories. From many such experiences those have purposely been chosen which lack more spectacular qualities and which may, for this reason, suggest answers to commonplace problems.

In the large general office of a factory, illuminated by good natural light from windows and skylights, there were persistent complaints of eyestrain and fatigue. The employees (mainly women using accounting machines) showed definite signs of restlessness and irritability and seemed frequently distracted from their work. Tests showed that the illumination was three times greater on a vertical than on a horizontal plane. Walls, painted in a light cream, were an evident source of glare and irritation.

The reactions on vision were as follows: The walls being brighter than the machines, constantly drew away the attention. In looking up, the general brightness of the interior tended to reduce the opening of the pupil. In directing the eyes again at work, there was a momentary halt period during which the employee had to wait for her eyes to readjust themselves. (The accommodation of vision is quick from dark to light, but slow from light to dark.)

When the walls were repainted in deeper tone, complaints largely ceased, and a noticeable improvement was shown in the efficiency and concentration of the office staff.

In a large, windowless drafting room, artificially illuminated with a good flood of fluorescent light, a change from white to soft bluish walls effected instant relief from eyestrain. In this instance, the visibility of grayish tracing cloth was improved, probably through the removal of distracting wall areas that previously had been brighter than the drawings and, therefore, of higher interest. The generally stuffy feeling of the interior (air-conditioned) was likewise overcome, and the normal temperature of the room (about 74°) was psychologically lowered.

In a hosiery mill, a topping operation involved the severe eye task of looping fine threads on a series of needles. The lighting conditions were apparently satisfactory. The employee, however, had to concentrate on a relatively small field, back of which was mere empty space. Brightness contrast was not only too great, but the outer fields of vision were too often exposed to the distracting movements of other workers.

Special backgrounds were constructed behind the machines, to block out the empty space. These screens, painted a light grayish blue, also reduced contrast and aided the eye in maintaining a comfortable adjustment. The result is well indicated in the report of management: "We have found that the experimental use of the blue backgrounds has worked out satisfactorily. The operators can do much better work with less eyestrain. It has not been possible to determine how much improvement has resulted from the use of this blue background in terms of increased production, but there is a definite improvement in that the operators of the topping equipment would not willingly do without this if their choice in the matter were considered."

From his own experience (and it has been rather extensive), the author is sure that color has an important bearing on efficiency and security. It may be that carelessness is to be blamed for some accidents; for poor workmanship, indifference, low morale. And you may be able to put your finger on this or that and say about it that color wouldn't mean a thing.

Yet color, intangible though it may seem, has real magic in it. Surely, abuse of the eyes may not be too remotely connected with the header that a worker may take down a flight of stairs. A sick stomach (resulting in a day off) may be due less to a salami sandwich than to the struggle of trying to see under conditions of excessive contrast or glare. A lost finger or hand on a guarded or an unguarded machine may, without too much imagination, be traced to a fatal indifference brought about by surroundings that might discourage any mortal, whether his forehead be high or low.

Again—Industrial Design

While this chapter may be somewhat irrelevant to the main subject of the book, the author feels that it will, at least, have some popular interest for the reader and perhaps add to his store of knowledge about color. After all, merchandise is made in factories, and the selling of color to consumers may not be too remote from the "selling" of color to industry, to effect the best of production and labor relations.

The engineering approach, however, should be of direct concern to the industrial designer. A large number of products are to be made better, more useful, and more beautiful when the color effect is purposeful, as well as appealing.

There are scientifically right colors for any number of materials and products. The designer, if he professes to be thorough and functional in his approach, should know that accurate methods and procedures are available to him in the matter of color and vision. He should read *The Science of Seeing* by Luckiesh, to learn how seeing specialists and ophthalmologists go about measuring visual efficiency and acuity. Perhaps he will recognize that such authority is helpful to call upon for advice and counsel.

A blackboard rightly designed in brightness and color would contribute to the well-being of a host of children.

A rightly designed automobile dashboard would affect many motorists.

Right colors in desk tops, in linoleum, in stair treads, in the handles of kitchen utensils, in baby buggies, in beach umbrellas (see the notes on heat reflection in the following chapter)—these and a variety of other commodities lend themselves to color engineering for the accomplishment of specific ends.

The functional approach, in fact, may often be used to supplement a purely aesthetic approach. A yellow tractor is not only appealing to the eye, but highly conspicuous as well. A light-colored doorknob is surely easier to find (in the dark) than is a black or a brown one, and the same is true of the shield around an electric-light switch.

Who can be sure that white sinks and black sewing machines are not causing trouble in innumerable households? Someone ought to apply the same research technique to a woman seeding grapes on a white porcelain surface as is applied in a factory for measuring the fatigue of workers in machine, assembly, and inspection operations. Industrial design—and merchandising—might find some new and potent sales arguments.

Chapter 12

COLOR ENTERPRISE—PLAIN AND FANCY

THIS is a chapter of miscellanea. It is concerned, in the first place, with a series of facts on certain physical properties in color, and it ends up with a general collection of strange uses to which color has been put by man.

Color lends itself to invention and ingenuity. It has been a hobby of the author to gather together various data, news items, and curiosities, all bearing on the anomalous. While such material may serve no direct purpose in problems of merchandising, it is interesting and it may, indirectly, inspire a bit of imaginative and resourceful thinking on the part of the reader.

Heat Radiation

When Piccard made his first jaunt into the stratosphere, the gondola of his balloon was painted black. The stratosphere has a constant temperature of 75°F. below zero. Yet the thermometer in his little ark registered over 100°F. above zero.

On his second flight, Piccard painted the gondola white. This time, the temperature dropped to a point below freezing. When Major Fordney and Captain Settle made their flight, they compromised by painting the upper half of the gondola white and the lower half black. Although a thermometer registered 103° above zero at the floor of the shell and zero at the ceiling, the general mean temperature inside was quite tolerable.

The sun emits heat. Its rays travel through space but do not get hot, so to speak, until they come in contact with something.

Because the black of Piccard's gondola caught the rays and absorbed them, the temperature naturally jumped. When white was used, it reflected most of the rays; therefore less heat was picked up.

One is thus able to explain the phenomenon of being able to ski with comfort, half naked, in the midst of winter. Though the actual temperature of the air may be below freezing, the skin is kept warm by the direct action of solar rays, which shower the body from the sky and from the snow-covered earth.

The technique of color in the control of temperature is very simple. White and all light colors repel heat through reflection; black and all dark colors absorb it. Some pigments, however, though dark to the eye, have special properties that make them efficient in infrared reflection. Paints with this quality, suitable for bus tops and storage tanks, are manufactured today by the Arco Company of Cleveland under the trade name of Infray.

The U. S. Bureau of Mines in its report No. 3138, gives some important facts on heat absorption and reflection as it affects evaporation in gasoline storage tanks of 12,000 gallon capacity. Over a period of 4½ months, a tank painted white and having an insulated housing had an evaporation loss of 112 gallons, or 1.40 per cent. A tank covered with aluminum foil had a loss of 170 gallons, or 2.12 per cent. A tank with aluminum paint had a loss of 187 gallons, or 2.34 per cent. A red tank had a loss of 284 gallons, or 3.54 per cent.

The difference of 114 gallons between white and red (and it would be larger still between white and black) would be serious in any sizable tank field. White, even though it cost more, would be economical.

More conservative facts on the efficiency of white in heat reflection are given by the Du Pont Company in a folder devoted to its Dulux White for Tanks. Here a saving of ⅓ per cent is mentioned, "in comparison with previous practical finishes." In quoting one case history, the average maximum temperature for a tank painted with Dulux white was 63.1° over a period of several months. For the other tank the average maximum temperature was 70.1°.

What this slight added efficiency would mean in dollars and cents is quoted as follows: "Assuming that white finish used on roof and shell of a 55,000 barrel cone roof tank has one-third of one per cent of normal tank capacity less evaporation loss per year than a tank finished with the paint previously considered best for heat reflecting properties. The tank containing 20,000 gallons regular grade motor fuel oil having a value of 0.5¢ per gallon. The saving would amount to 6,600 gallons per year, or $330.00. With the application of both finishes at the same hourly rate and one-third more time being required to apply a 'double-coat' of white, the evaporation saving would offset the higher applied cost of Dulux White for Tanks in about 6 months. A saving would result through the use of Dulux White for Tanks of $150.00 the first year and $330.00 per year, for several years before the need of repainting."

As to home radiators, aluminum and gold paints have long been thought most efficient. The fact, however, is that metallic paints reduce heat transmission and are not so desirable as other types.

Heat is distributed from a metal surface through convection (the air that passes over the radiator) and through radiation (the rays that are sent through space). Aluminum paint and dark colors may produce a higher surface temperature on the radiator, but they may also trap radiant energy from escaping. White or light paint lets these heat rays through by transmission. The result is that more heat is distributed throughout an interior and that there is more warmth in distant objects and surfaces that catch the rays and convert them into heat. So-called radiant heating systems, employing long lines of pipes in floors or walls and depending more on heat by radiation than by convection, are today fast coming into use.

Regarding the relative merits of metallic and light-colored paints on radiators, the U. S. Bureau of Standards in its Letter Circular 263 has this to say: "The effect of adding metallic paint is equivalent to removing ⅙ of the radiator or nearly 17 per cent. Or as if one section out of six had been removed. Thus a radiator of five sections painted with white or a light color should be about as efficient as another of six sections painted with metallic paint."

Other examples along this line and other enterprises are to be quoted. Aluminum foil is used on chocolate bars to reflect heat and thus prevent high temperatures from melting or discoloring the candy. It has also been extensively employed for the insulation of refrigerators, ships, airplanes, homes. Its action, however, is one way. That is, it resists heat but not cold; yet, coupled with thick, fibrous layers of other insulating materials, it is quite effective.

From the functional standpoint, therefore, white and light colors have almost endless applications. A white ship in the tropics will be at least 10 degrees cooler on the inside than will a black ship. Incidentally, light colors seem to reduce the growth of barnacles. Barnacles apparently dislike brightness and prefer to snuggle up on the dark undersides of a ship. Light-colored hats, clothing, tents, awnings are wise summer measures. Baby carriages certainly should not be black or dark gray.

Automobile tops, airplane cabins, water tanks, freight cars, trucks, and so on and on, should advisedly take advantage of the heat-reflecting qualities of color. The heated and air-conditioned colonial home should be white (as it generally is); but the unheated barn, in real need of the sun's energy on a cold winter day, should be red (as it generally is).

Black paint has been used on ice to hasten spring melting. Small harbors and rivers have actually been opened to traffic well ahead of nature's own schedule through this device. From Russia comes the report of a rather amazing project. Here consideration has been given to the plan of sprinkling coal dust on the snow-covered slopes of the Ukrainian mountains to hasten spring thaw.

Ultraviolet Radiation

Ultraviolet radiation tans the skin, is responsible for vitamin D, fades colored merchandise, and raises havoc with many food products. Here, again, color offers an effective control.

While ultraviolet has little power of penetration, certain plastic sheets are fabricated by industry today to transmit its energy, for windows in a chicken house, for example, where the rays are beneficial. Conversely, other window glass is manufactured to screen it out. A glass called Solex not only resists ultraviolet, but also is efficient in rejecting the thermal qualities of sunlight. It is, therefore, appropriate for use in hot climates.

Brown bottles for Schlitz represent another functional application of color to thwart the more evil action of light. On the market today will be found amber Cellophane, specifically made to admit clear visibility of certain food products but to deny access to harmful rays. Total opacity, of course, will accomplish the same result. Further, light colors on packages will resist heat and thus help to prevent a too-rapid deterioration of their contents.

Where the product is to be seen and is, therefore, placed in a transparent container, some hues are superior to others for this purpose. In a study conducted by the Food Research Division of the U. S. Department of Agriculture, these facts are mentioned: "Light has long been known to catalyze this form of spoilage [rancidity in oil-bearing foods], but not until recently has it been shown that certain wave lengths of light promote rancidity more than others do. Investigations have been conducted by the Bureau of Chemistry and Soils with various color filters chosen to absorb known wave lengths of light, so that all regions of the spectrum were used selectively for radiation. . . . " The ultraviolet portion of the spectrum catalyzes rancidity the most, while the violet, "indigo, and blue of the visible spectrum appear to be next in activity. Yellow, orange, and red are also active in producing spoilage by rancidity, but about twice the amount of irradiation with these wave lengths is necessary to produce the same amount of change or spoilage as with blue. The green region of the visible spectrum and the infrared of the invisible appear to be practically inert to rancidity development. Oil-bearing foods are, of course, best preserved from the development of rancidity when protected from all light. Our experience shows that a green filter transmitting light between 4900–5800 A.u. [a yellowish green] is next in protective properties."

Apparently, there is need for further research—for instance, on the deterioration of vitamins. An item in the *American Journal of Pharmacy* points out that "The substitution of glass jars for cans may well result in destruction of the riboflavin and even ascorbic acid in preserved fruits and vegetables unless the glass container is protected from light."

Fluorescent Colors

A considerable business, these days, has been built upon the use of so-called black light (ultraviolet) with fluorescent materials, pigments, dyes, inks. Fluorescence is that phenomenon in which light energy of one frequency is converted into energy of another frequency through the curious action of certain substances.

There is a product called Soilax, for example, which in its dry state is a pink powder to be used as a washing compound. Soilax contains a fluorescent substance, which has been thoughtfully added to "write" visible instructions when the product is mixed with water. Thus, when a little water is added to Soilax, the result is a yellowish solution. However, when just the right amount of water is used, the solution is green. To quote from an advertisement, "Ten minutes will instruct your employees in the correct use of Soilax; the least intelligent can't fail to understand and heed its warning color signals."

With ultraviolet radiation, this magic of fluorescence becomes spectacular indeed; for, while the energy of the light source is itself invisible, the fluorescent materials it activates shine forth out of darkness in brilliant, luminous hues. Theater carpets, special murals, and displays, all treated with fluorescent pigments and showered with invisible ultraviolet rays, loom before the eyes to offer exciting visual experiences.

Fluorescent materials and ultraviolet light combine to serve a number of useful purposes and to inspire many strange color enterprises. Those uses mentioned here are selected from data assembled by the Continental Lithograph Corporation of Cleveland, which, through its Conti-Glo Division, is perhaps the largest purveyor of fluorescent products and equipment in America.

First of all, a number of dramatic uses are to be found for black light in theater, hotel, restaurant, and store. These offer great artistic and commercial possibilities in the never-ending quest for novelty with color, for effects and illusions that come to the public as a startling and original surprise.

Other applications, perhaps a bit more resourceful and practical, have been developed and reviewed by the Conti-

Glo organization. When maps or printed matter are dusted with a fluorescent powder, these can be read under ultraviolet light, even though the room itself may be in total darkness. To catch the vicious or petty thief, money, cash registers, valuable tools, or other possessions may be sprinkled with fluorescent powder. When these are handled, the powder catches upon the clothes or the fingers of the culprit. Though he may see nothing, his guilt will be revealed when, on his way out of the factory, house, or store, he passes under the rays of an ultraviolet lamp. Instantly, the powder will fluoresce to a vivid green and the infamous secret will be out!

Because different chemicals and substances have different appearances under visible light and ultraviolet light, much detective work is possible. Forgeries, alterations in written documents, counterfeit materials may, through fluorescence, make themselves known. And such analysis may be for good as well as evil.

Oleomargarine added to butter will show blue fluorescence. Wool may be distinguished from cotton. Beet sugar shows little fluorescence; cane sugar and glucose give off a reddish light. False teeth, freckles, scars, diseased tissues show up when activated by ultraviolet.

Hidden flaws, invisible marks, spots, stains shine forth in many materials. "The ultraviolet differentiates many apparently identical materials, structures, compositions, elements, apparently identical fibers, compounds, coatings, solids, liquids, gases, and the constituents of mixtures which are absolutely uniform in daylight."

The science of medicine, as well, has taken ingenious advantage of fluorescence. *The Reader's Digest* has recently reported on an item that appeared in *Medical Clinics of North America.* To determine whether or not a diseased or an injured part of the body is receiving a proper supply of blood, a small amount of a nonpoisonous dye, called fluorescein, is injected into the blood stream. This dye shows green under ultraviolet light.

"Drs. Lange and Boyd have applied the 'green-light test' in many demonstrations at the Flower and Fifth Avenue Hospitals in New York. A patient with a possibly gangrenous leg is wheeled into a darkened room and given an injection of fluorescein. If at the end of 20 seconds the leg glows golden-green under an ultraviolet lamp, circulation is sound and surgery is unnecessary; if the leg remains dark, then it is gangrenous and must be amputated. Furthermore, by observation where the green glow of circulation ends, the surgeon can select the ideal site for amputation."

Concerning Better Mousetraps

For strange exploitations of color the effort does not always have to be complex and scientific. The merchandising man who decided that a Scotch-plaid horse blanket would be more glamorous than an olive-drab one had a wise and happy thought, for which he was duly rewarded in increased sales. The same is true of the maker of garden implements who found that red handles produced quicker sales, the extra cost being little and the extra utility none. It was a cannery packer of fish who, when he had a run of white salmon, labeled his can so as to give assurance that "This salmon is guaranteed not to turn pink." Thus he prepared his customers for the light color and perhaps insinuated the disturbing thought that other salmon *turned* pink.

On the other hand, the baker who concluded that brightly hued bread—pink, green, blue—would make attractive sandwiches was not so wise. Here, of course, the experiment was destined for failure because of the innate unwillingness of human beings to stomach colors that appear psychologically inedible.

Another figure on the positive side is the maker of coated candies who found that a variety of colors sold more pounds, even though the flavor was the same. Still another example is that of the progressive oil company that extended its activities into India. Luckiesh in his book *Color and Colors,* writes: "Oil was sold in tin cans for use in lamps. On the rectangular faces of the tin a red monkey was printed. Apparently, a consular or company representative knew that the natives held the monkey in high esteem, and of course they liked red. The sides of the tins were hung on the walls of the humble huts by the natives for their own enjoyment and reverence. It cost no more to print the red monkey on the sides of the tins than the trade-mark used in this country."

A few more instances of clever enterprise will be reported here, not only for the sake of interest, but in the hope that these strategies and inventions may encourage equally shrewd thinking on the part of some readers.

Color and War

To Egmont Arens of New York we are indebted for a collection of notes on the use of color during the war. Arens writes: "Using colors for identification and signalling is not new in itself. What is new is the scientific understanding of human reactions to color, and the use of color *dynamically* to get faster action at a time when speed is the essential ingredient. . . . On training field and battle front, on battleship and bomber, color stands by to organize, simplify, and direct *action*. In fact, wherever the human eye is apt to err, or the human hand to slow down with fatigue or confusion new scientific color applications have been discovered to simplify complex operations, to heighten speed and precision of action, to lessen strain and tension, increase work output."

In an airplane, the eyepiece of the gun sight is red, to aid instant location. On instrument boards, colored lights and marks distinguish functions, warn of danger, provide constant visual check. Electric wires and oil lines are coded by hue, to facilitate quick repair in the event of damage. Deck crews on aircraft carriers wear bright-colored caps and sweaters so that they will be conspicuous to landing planes. Training planes are brilliant yellow or orange, to help in preventing crack-ups among neophites.

Ammunition is marked in color, bomb casings being thus identified as to type—incendiary, fragmentation, armor-piercing. Likewise, torpedoes, depth bombs, machine-gun bullets, and tracer bullets have separate color symbols to facilitate proper loading of cartridge clips.

The insides of tanks used in desert warfare are painted pale green, both to reflect light and to offer psychological compensation for exposure to high temperature. Colored smoke signals, which may be changed from day to day, are used for tank identification and for preventing confusion (or trickery on the part of the enemy). The undersides of self-inflating life rafts are so colored as to make them less attractive to sharks.

A multipurpose accessory fabric is yellow on one side, to command notice, and blue on the other side for concealment. A bottle of fluid is available to stain the ocean and hence give the rescuing aviator a larger area for his eyes to seize upon.

Incidentally, in the old days, Lord Nelson is said to have painted gun carriages and gun decks red—to keep his warriors from becoming unnerved by the sight of bloodstains.

Of Flies and Mosquitoes

The war has stimulated, among other things, a further inquiry into color reaction among insects. The results given below are obtained from notes assembled by Deane B. Judd and presented in *News Letter* 45 of the *Inter-Society Color Council.*

For house flies, several investigations have led to contradictory results. E. Hardy, for example, found yellow to be avoided and white to be preferred. On the other hand, P. R. Awati considered yellow to have the greatest attraction, red and violet the least. O. C. Lodge found no preference at all. S. B. Freeborn and L. J. Perry found the fly repelled by pale colors, while R. Newstead had reason to conclude that light colors were preferred to dark colors. (Something must be wrong somewhere. Either the methods of research are unreliable, or the flies with haunts in different parts of the world have different ideas about the matter.) Possibly, the safest conclusion is that flies are more attracted to lightness than to darkness, for the weight of evidence seems to indicate as much.

In Holland, at least, horse stables and cow stalls are frequently treated with blue to get rid of the pests. So in Holland the flies must dislike that hue.

Regarding mosquitoes, however, the authorities are in far better agreement. Here light colors are the repelling ones. G. H. F. Nuttall and A. E. Shipley found that the common European malaria-bearing mosquito alighted most on dark blue, red, and brown, and least on yellow, orange, and white. (Subsequent to this particular report, the U. S. Army withdrew its regulation navy blue shirts in malaria districts and substituted lighter colors.) Shariff, similarly, found during 5 years in South Africa that pink and yellow mosquito curtains did not harbor insects. When boxes were lined with navy blue, pink, gray, and yellow flannel, the interiors of the blue and gray boxes were thickly covered with mosquitoes, while but two or three were found in the pink or the yellow boxes. Hoodless also found that New Caledonia mosquitoes prefer blue and avoid yellow.

Of Chicks, Eggs, and Cotton

Now let us turn to a few items and clippings on widely different topics.

From the New York *Herald Tribune* comes this: "The mystifying spectacle of 3,000 white Leghorn chickens at the Essex County penitentiary's poultry farm wearing red goggles was explained today by Warden Floyd Hamma, who said that the rose-colored glasses made them more peaceful and had stopped the fighting which had resulted in the death of 10 per cent of his Leghorns this year.

"Matthew Gaynor and Fred Kirsch, the guards in charge of the poultry farm, which supplies eggs for other institutions of the country, found that among the 8,000 chickens which they keep, the killing among the Leghorns was disastrously high. They found that a harmless peck causing a slight scratch immediately was transformed into a death fight by the sight of blood.

"If the chickens saw red all the time, the keepers thought, this felonious tendency might be curbed. Consequently, the windows of the chicken houses were tinted scarlet, but the killings continued. The keepers and the warden conferred on the matter a month ago and emerged with the plan to use red goggles. They devised goggles of a small strip of leather fitted with red isinglass and wired to the beak. This devise first was tried on 100 chickens, and at present 3,000 have been equipped. Since the red goggles were put into use, the warden said, there has not been a fatal fight among the Leghorns."

A similar protection has also been instituted in a Chicago "broiler factory," where chickens are raised like so many mechanical products amid artificial sun lamps, endless belts, and special diets. Cannibalism brought about in chicks at the sight of blood when pin feathers push through the skin has been curbed by "washing out" clear visibility of the color with appropriate red lights.

A group of scientists at the Kansas State College have succeeded in getting hens to lay red and green eggs and eggs with snow-white yolks. To quote from a newspaper report: "Production of eggs of practically any hue can be accomplished by simply feeding the proper organic dye in each case. At present, the most fascinating problem with hens is to determine the exact precursor of the pigment astacene, which is synthesized by the chicken. This pigment is found in only one part of the chicken's body, the retina of the eye, and is present in no normal feed ever tested. There is evidence that this pigment may play an important role in the color vision of the fowl."

From San Diego (*Southern Rancher*) comes a lengthy report on controlling the color of meat birds and eggs. "The color of the fat and flesh of meat birds is extremely important in many markets as the particular type of trade demands milk-white flesh, skin, and fat. There are, however, other sections and nationalities that have no objections to skin and fat ranging from light yellow to dark yellow. The white-skinned and white-fat birds are marketed as 'milk fed' and bring a premium over the dark. The term 'milk fed' is more a trade name than a fact and could be more aptly called 'controlled color.'

"The same situation has existed for many years in the New York City egg markets. The New York egg market demands an egg having a lemon yellow yolk and they will not accept eggs with darker colored yolks. As most of the Pacific Coast surplus eggs are shipped to the East, they are candled and the darker colored yolk eggs are sold at home and the lighter colored ones shipped East. The color of the yolks affects grading and generally the lighter colored yolks bring better prices than dark ones. . . .

"Now the problem of controlling the color of egg yolks and the color of meat birds in order to obtain the best market values has been studied for many years and the results of some of this work follows:

"Dr. W. A. Maw, head of the Department of Poultry Husbandry, MacDonald College, Quebec, Canada, gives the results of studies that have been carried on for ten years.

"'The color of the body fats and skin is closely related to the color of the cereals in the ration fed. Where yellow corn is fed to take advantage of the additional body fat produced, the color of the fats is too yellow to classify as milk fed for market purposes. A study was therefore made of the bleaching effect of such agents as Bone char, English nut charcoal, etc. Yellow ground corn was used as a cereal base in the rations. The differences in the color of the fats and the skin were determined on the basis of the amount of carotinoid pigment in the fats. The bleaching agents reduced the amount of carotinoid pigment as compared to the amount shown in the basal ration fed fats. Bone char showed the best results, producing relatively white-colored fats. English nut charcoal did almost as good a work.' . . .

"Tests have recently been completed by several California milling firms to determine the possibility of the control of color in the egg, meat, fats, and skin. These tests have been conducted both on controlled battery birds and under flock-size conditions. The unanimous opinion is that 'pigmentation of the skin, fat, and color of the egg yolk can largely be controlled by varying the amount of charcoal fed.'"

This control over nature may also, in the future, be extended in other and unusual ways. The problem of color styling in cotton textiles, for example, may one day require a thorough knowledge of horticulture. Colored fabrics may be grown from the seed, obviating the need for costly dyeing processes and, at the same time, letting nature handle the responsibility of permanence and resistance to fading. *The New York Times* has this to report: "Naturally colored cotton from which fade-resistant cloth is woven is being grown in Russia. . . . Thus far, green, rose, lemon, and brown hues have been developed. Last year 1200 yards of cloth were woven from the colored cotton, and this year 700 tons will be grown from which 1,000,000 yards of colored cloth will be produced."

Of Tomatoes, Corn, Peaches, and Foxes

When is a tomato ripe? Some clever thinking at Purdue has turned cosmetics to the purpose of color standardization. Thus states the *Food Field Reporter:* "Tomato pickers all over the country have been having trouble picking fully ripe tomatoes, because they have had no way of telling how red the fruit really was. So the Agricultural Department of Purdue University got busy and turned out a nail polish that is the shade desirable in ripe tomatoes. With this polish on their fingernails, all the pickers have to do is find tomatoes as red as the polish, or redder. This is expected to aid farmers in raising their inspection grades.

"The College Inn Food Products Co., Chicago, was quick to make use of this novel idea. It provided each of its pickers with a bottle of nail polish, and expects the quality of its tomatoes to rise immediately."

From the Kansas City *Star* comes notes on colorless plants: "Albino animals are interesting freaks, often valuable because of their rarity. Albino plants beyond the seedling stage are practically unknown, because they automatically starve to death, lacking as they do the green pigment, chlorophyll, necessary for the manufacture of basic foods.

"Physiologists of the Carnegie Institution at Washington, however, under the leadership of Dr. H. A. Spoehr, have been able to keep albino corn plants alive for four months and more by feeding them on sugar and other nutrients. In this way they have begun to obtain an understanding of some aspects of plant nutrition hitherto inaccessible. It has been discovered, for example, that such plants can manufacture plenty of starch if supplied with ordinary cane sugar, a process hitherto considered impossible. It has also been found that they cannot get adequate nutrition out of glycerine and other compounds, which have until now been looked upon as adequate plant foods.

"Parallel experiments have been carried on with plants artificially albinized by putting them into the dark while still young. Sunflowers so kept on a diet of sugar and the necessary minerals have grown and even blossomed, producing normal yellow-rayed flowers although their leaves wholly lacked green color."

Another strange phenomenon regarding color and foods is given in the *Los Angeles Cultivator:* "Proof that more than normal color can be brought out in peaches and nectarines by exposing them to methyl bromide gas has been obtained at the College of Agriculture by Dr. L. L. Claypool, assistant professor of pomology.

"Unfortunately, says Dr. Claypool, methyl bromide cannot be used commercially to produce more colorful fruit. For while it increases the color, and therefore the eye appeal, of peaches and nectarines, it also damages their flavor.

"The value of the discovery, says Dr. Claypool, lies in the proof that the color of some fruits can be improved through exposure to a gas. The problem now is to find a gas which will cause this improvement without damaging the flavor.

"He points out that the gas treatment will work only on fruits which already have a tendence to 'blush.' He believes the increased color produced in peaches and nectarines may be due to a reaction between methyl bromide and pigments present in the fruit in colorless form."

And now an item about the vertebrate fox from the Easton (Pa.) *Express:* "Because of the preference for platinum blondes, the silver fox is put on a vitamin B deficiency diet.

"This diet, claimed Dr. Agnes Fay Morgan of the University of California, Sacramento, artificially produces platinum pelts from silver foxes, similar to those imported from Norway last year, which proved so popular and brought extraordinary prices."

Of Various Other Things

From another newspaper report, on a radically different matter, comes the following: "A means of detecting poisonous gas was indicated by a new device developed by T. T. Woodson, of the General Electric Company. The apparatus works on the principle that a vapor will absorb certain colors; for example, mercury absorbs violet.

"A violet light is passed through a sample of gas and on to a photoelectric cell. If mercury vapor is present, part of the violet light is absorbed and the decrease in light transmitted is detectable by the photoelectric cell, which actuates a warning signal."

From the New York *Daily Mirror* we quote the following: "Boric Acid Kills 5th Baby; City's Hospitals to Color Poison Pink. Death claimed another infant victim of New London's boric acid mixup yesterday, bringing the toll of that tragedy to five, and New York Commissioner of Hospitals Bernecker took a final, decisive step to insure against such a thing ever happening here.

"Louis Samuels, chief pharmacist for the city's 28 hospitals, was ordered to instruct all municipal pharmacists to color boric acid a bright pink, through use of amaranth, a harmless synthetic used in the past for foods, drugs, and cosmetics."

From the Clayton (Ill.) *Enterprise* this article is taken: "Did you know that colors can be warm or cold, that some colors absorb and radiate heat? A Californian expert, W. F. Alder, has produced a new paint which absorbs great heat, and then radiates it. He tested two steam-heated cubes of polished copper, one of which was painted with his compound. This was found to have a temperature of 40 degrees Centigrade more than the unpainted cube.

"Trunks of fruit trees were painted with this mixture, and during a spell of severe weather they defied frost. So in the near future we may be able to partially heat our homes with paint."

Another new development is reported from London: "The possibility of revolutionizing interior lighting is foreseen by the announcement here today of a new luminous paint discovered by chemists. . . . Known as 'lumogen,' this material can be mixed with paint of any color and will then glow brightly when subjected to ultraviolet rays.

"The chief advantages of 'lumogen' are said to be its low cost and its excellent luminescent qualities under all conditions. It can be mixed with almost anything from concrete to wallpaper or floor polish, so it will be possible to create rooms with glowing walls, ceiling, and floor, and even a luminous Broadway."

The following comes from Fresno, Calif.: "One of the latest scientific advances for measuring strain on automobile parts is to make a transparent plastic model of some highly stressed metal part, such as a gear or a connecting rod, and to examine the model by passing polarized light through it. When the model is subjected to loads which represent the conditions of actual use, bright bands of color appear, like miniature rainbows, indicating the sections of greatest strain."

To facilitate instruction in the intricacies of blind flying, a novel use of red and green has been reported in *Illuminating*

Engineering: "Blind flying equipment is a convenient new method of fitting standard planes for safer, more effective blind flying practice in daylight. The equipment consists of transparent colored sheeting that is attached to cockpit windows, and transparent colored goggles worn by the student pilot. Wearing the goggles, the student pilot can read his instruments, but cannot see outside the cockpit. The instructor, who does not wear the goggles, has an unimpeded view of ground, sky, other planes, instruments and pilot. . . .

"The combination of goggles and sheeting acts as an effective one-man blackout system because of the remarkable color selectivity of the two plastics used. The green sheeting transmits only green light, blocking all other colors. The red goggles are fitted with a lens which transmits only red light. The red and the green colors are precisely formulated and carefully controlled, so as to be mutually exclusive. This means that while it is easy to see through either the sheeting or the goggles separately, it is impossible to see through the combination. Consequently, the student pilot wearing red goggles cannot see through the green sheeting which covers the windows of the cockpit, although he can see everything within the cockpit. The instructor, not wearing the goggles, is able to see clearly through the green sheeting as well as within the plane."

The Covington (Ky.) *Enquirer* offers this account: "When called upon to paint an ice skating rink for a carnival, a St. Louis contractor found it could be satisfactorily done with a spray gun. After the arena had been flooded with an inch and a half of water and frozen solid, a special ice paint colored blue was sprayed over the ice. It froze almost immediately, drying with a dull surface like flat wall paint. To give a varnished look, water was sprayed over the paint coating to allow ice one-sixteenth of an inch to form. About 125 pounds of the special ice paint in powdered form was used for the rink, which measured 108 by 215 feet. It is said that skate runners do not change the color of the finish because the tracks refill with the color. Ice may be painted in more than one color, or in patterns, but a brush cannot be used to apply it, as the paint freezes so rapidly it will leave streaks."

Of Mental Telepathy

This is color enterprise. To keep up with all of it, a person would need many assistants and the help of several professional clipping services. Man, forever fascinated by the spectrum, puts it to more and more uses and services. Not alone in his great laboratories, but in the simplest walks of life, as well; for instance, the resourceful shepherd who trains one black sheep to accompany 150 ewes. He has only to stand on a high place and count his black sheep to know if his whole flock is intact.

Finally, a quotation from an article in the New York *Herald Tribune* will lead the reader on to the rather fascinating subject of the next chapter.

"A telepathic antagonism to the color red has been revealed by experiments in extra-sensory perception conducted by Professor Gardner Murphy and Ernest Taves, of the department of psychology, Columbia University, as part of the research program of the American Society for Psychical Research. . . .

"In many of the experiments the choice was limited to two possibilities—yes-no, heads-tails, black-white. Other tests employed the ordinary extra-sensory-perception cards and a deck called 'Rook cards,' fifty-six cards made up of four suits, each suit numbered from one to fourteen and each suit in a different color. . . .

"When the readings of the Rook cards were tabulated and analyzed it was found that the group as a whole had a slightly above-average record. When the results were subdivided by the color of the cards it was found that when calls were made of the suit in which the figures were printed in red on a white background the scoring dropped significantly lower than in any other test. . . .

"This effect involving red was considered so interesting that Professor Murphy communicated with several professors of psychology in universities throughout the country asking that the test be repeated by them. He made more extensive tests with the red cards among the American Society for Psychical Research group, running the total readings up to 6,975. Again the antagonism to red was found. . . .

"Despite the fact that the subjects were not informed when they were working with red cards, they became aware of it. Somewhere below the level of consciousness they sensed that involved was some faculty which prevented them from making a record as good as usual in identifying the cards by telepathy. This faculty had the same effect on the telepathic process as a red traffic light or a red danger sign. . . .

"The Columbia scientists, in commenting on the unusually low record on the red cards, declare: 'We have tentatively adopted a hypothesis, but further research is necessary before much can be said about it. It would appear, at least, that there may be something about the red-white situation which arouses negativism. This may be because of obscure effective factors—the symbolic values of red and white are deeply ingrained in our culture, at least, and this general effective tone of the material used may have something to do with the type of results obtained.'"

Chapter 13

THE PSYCHOLOGY OF COLOR

THE author hopes that the reader will understand the purpose of this and the next chapter. They are concerned less with practical instruction than with a review of odd and sentimental things more related to the background of color than to its real substance. However, just as a doctor should be interested in the history of medicine, or a lawyer in the beginnings of law, those who work with color should find interest and profit in the romance of their own endeavor.

Many of us are frequently called upon to popularize the subject of color, to design booklets, releases, and promotional material that will excite public fancy and thus enhance the products or services that are the more practical concern of business. A wellspring of such reference material will be found in the pages that follow.

There is, for example, value in bringing together all available data on the so-called psychology of color. While the term is not easy to define, most of us, in using it, have in mind the influence and affective power of color with reference to the mind, emotions, and body of man.

Incidentally, this makes for good reading, just as it adds authority and knowledge to all efforts at selling color. Reviews of the psychology of color are hard to find. Few writers on color have gone to the trouble of digging into scientific and medical literature, sometimes for a page or two out of an entire book, or for a few sentences lost in a maze of phraseology that is frequently more confusing than a foreign language.

This present chapter, the author admits, has been a difficult one to gather together; yet he feels that the inconvenience of consulting medical libraries, technical books, monographs, publications, and reports, to satisfy his own inquisitiveness, has been more than balanced by the pleasure he has had in it and by the prospect of satisfying some of the natural curiosity of others.

Color and Healing

The influences of color on man's being probably need no strong defense. Since the beginning of time he has looked upon light as the emanation of a divine and omnipotent force. In this he has been less fanciful than practical, for color has always been associated with life itself. To the ancient the hues of the rainbow were symbols, which he related to almost every aspect of his civilization. Far more of a mystic than an artist, he was convinced that color ruled over his destiny and security. The same impulses that all of us experience today in the sensation of color were to him the whisperings of high powers, and he promptly sought to comprehend them and turn them to good purposes.

In the art of medicine the Egyptians diagnosed and healed with color. Papyri dating back to 1500 B.C. prescribe medicines compounded of certain colored materials. Pythagoras in the sixth century B.C. healed with music, poetry, and color. Celsus and Galen in the first century A.D. wrote of yellow and red poultices: "There is one plaster almost of a red color, which seems to bring wounds very rapidly to cicatrize."

Avicenna, the Arabian, in the Dark Ages, declared that red and yellow were injurious to the eyes, blue soothed the movement of the blood, red stimulated it. Paracelsus and the alchemists, too, were all convinced that the philosopher's stone and the elixir of life (invariably red) were within the ability of man to formulate. Health was a condition of harmony with divine forces; disease, a disruption of harmony. Colors, invocations, talismans, herbs, leeches, must be called upon to counteract affliction.

The Modern Attitude

After many centuries, this viewpoint was suddenly interrupted by the microbe hunters. With the mysteries of disease exposed in the microscope, men abandoned all things presumably superstitious and occult, and for a long while the therapy of color was forgotten.

The modern revival came toward the end of the nineteenth century in America. S. Pancoast, in 1877, called attention to the favorable influence of certain colors upon the growth of plants and the health of animals. A year later, Edwin D. Babbitt published his *Principles of Light and Color*, which ran into many editions, was translated into several languages, and brought to him the rather ecstatic reputation of a miracle man. Regardless of Babbitt's integrity, he did succeed in shocking the medical profession out of its stubborn apathy. He is the man who adjured prudent Victorians to expose their nakedness to the sun, who created fads for colored-glass alcoves and colored-glass lighting fixtures. Years before the invention of ultraviolet lamps, he recognized the efficacy of light and, according to case histories, which probably might be questioned, he cured a multitude of diseases.

The modern physician, however, holds to a skeptical attitude. He admits the power of ultraviolet, for example, when he can look into a microscope and see germs destroyed. But he cannot, with equal certainty, believe that the *visible* colors of the spectrum also are potent.

Despite much doubt, the art of healing with color gains steadily in prestige and recognition. Years ago, Picton noted the effect of light and color in cases of smallpox. Niels Finsen, one of the earliest exponents of light therapy, observed that skin ulceration did not occur under red light. Here, however, red offers a sort of reverse therapy. Light in general is harmful to the skin of the patient afflicted with smallpox; while red seems to be negative, and therefore beneficial.

The author's notes on the therapy of color in the treatment of disease are not, perhaps, to be taken too seriously. They have been assembled from medical literature, mainly from work done in England, and will give some idea as to an efficacy on the part of color that may some day be more actively a part of medical practice.

Red light has been used to treat erysipelas, uticaria, scarlet fever, measles, eczema. Red and infrared light are being used to mitigate the burn produced by ultraviolet light, and to treat fractures. Because red produces heat in the tissues and dilates the blood vessels, R. Douglas Howat, in articles for the *British Journal of Physical Medicine*, has referred to its value in lumbago, arthritis, and neuritis.

Blue, on the other hand, has bactericidal properties. It is said to increase the output of carbon dioxide in certain cold-blooded creatures, whereas red exerts the most influence among birds and rodents. Moleschott has shown that light in general increases the elimination of carbonic acid in animals. Blue has been prescribed to cure "thumping" headaches, high blood pressure of nervous origin, and intractable insomnia. Whether its action is direct or indirect (through the eye and emotions) is something to be determined more surely.

Yellow has been found by some qualified investigators to raise blood pressure associated with anemia, neurasthenia, and general debility. Applied abdominally, it is said to increase the flow of gastric juices.

Green seems to be generally neutral.

In other experiments, a definite antagonism has been found between red light and ultraviolet light. Substances activated by ultraviolet light have been rendered inactive by red light. The activity of hormones has been increased by red and destroyed by ultraviolet. Ludwig and von Ries believe that various portions of the spectrum have a specific influence on the hormones of the body, and that endocrine problems may some day find their solution through more intensive photobiologic studies.

Modern medicine may hesitate to accept color for its direct therapy, but certainly makes common use of it in diagnosis. Some years ago Dr. John Benson, in an article for the *American Journal of Clinical Medicine* (December, 1907), remarked that as a general rule a white tongue reveals a system in need of alkalies, while a person with a bright red tongue is likely to need acids. He pointed out that dark red is often a sign of infection, or sepsis, a brownish red tongue a sign of typhoid. The broad, flat, flabby tongue may indicate a gastric intestinal wrong.

Pigmentation of the skin may be even more symbolic. Benson presented the following interesting list:

Thick, dirty, muddy complexion (also the ruddy face and bulbous nose): evidence of autotoxemia.

Yellow face: hepatic condition.

Greenish, waxy skin and pallid lips: a sign of anemia.

Restricted spots of deepening red, often shading into purple: pulmonary lesions of pneumonia.

In some diseases the afflicted person may experience colored vision, a sensation in which the field of view appears weakly or strongly tinted. In jaundice the world may appear predominantly yellowish. Red vision may follow retinal hemorrhage or snow blindness. Yellow vision may follow digitalis or quinine poisoning. Green vision may be caused by wounds of the cornea. Blue vision has been reported in cases of alcoholism. In tobacco scotoma the vision may be reddish or greenish.

In santonin poisoning the world may at first appear bluish. There may be a second state of yellow vision of longer duration and a stage of violet sight before complete recovery. Following the extraction of a cataract, the patient may experience red vision, sometimes followed by blue vision.

Physical Reactions

Colors have effect upon the mind, emotions, and body of man. They cure diseases, indirectly if not directly. They have influence upon human efficiency. The sensation of color is affected by the condition of the body, and the body in turn is affected by colors. They are important to human welfare, and they offer therapies that are of real consequence to the individual and to society. Luckiesh writes, "One should not be surprised if it is revealed some time in the future when we know more about the human being, that all wave lengths of radiant energy from the sun and sky are intricately entwined in the life and health processes of human beings."

In 1910 Stein called attention to a general light-tonus in the muscular reactions of the body. The word tonus refers to the condition of steady activity maintained in a living person. Muscular tension and muscular relaxation, for example, are tonus changes. Feré discovered that all light increases tonus, blue slightly from a normal 23 to 24, red considerably from a normal 23 to 42. Metzer observed that when light was thrown upon one eye a tonus condition could be produced in the corresponding half of the body. Daitsch and Kogan, following a series of experiments, concluded that yellow and purple light had the best effect upon human metabolism. Red tended to weaken it considerably, green to weaken it slightly.

Even the human aura is being studied. Bagnall of England has mentioned its possible value in the diagnosis of such afflictions as epilepsy. Each of us, of course, has an aura—heat rays (as well as odors) emanating from the body. Some years ago, Vanderplank announced that the tawny owl had vision for infrared radiation and could find its prey in what to man is "pitch darkness" by actually seeing heat as it emanated from a mouse or a rat.

An interesting series of experiments with animals and colors was conducted a few years ago in Japan. These experiments, which are well worth mentioning, will be found reprinted in the *Japanese Journal of Obstetrics and Gynecology* (1940).

The secretion of milk in guinea pigs was accelerated when the mammary glands of the animal were irradiated with red light. Blue rays retarded the secretion. The young of those mothers which had been exposed to red light grew strong and healthy, while the offspring of the blue-irradiated mothers were undernourished. Dr. Menju who conducted the tests declares, "I am convinced that these effects of the visible light upon the milk secretive function are induced through the vegetative nervous system and the function of endocrine glands"

In the study of the sex cycles of animals, blue light was found to cause irregular cycles. Red light caused shorter and regular sex cycles. "Long wave length and short wave length have antagonistic effects upon the sexual cycle, thyroid gland, ovary, and adrenal body." Similar results have been achieved in America by Bissonnette with starlings, the sexual cycle being hurried along in male birds by the action of red light.

As to the general effect of color on the vegetative nervous system of rabbits, Dr. Yogo of Japan concluded, "In case red ray radiation is applied to the whole body or abdomen, it affects to accelerate the tension of the parasympathetic nervous system." Its first action is to increase blood pressure. After irradiation, however, the blood pressure may grow steadily lower.

"In case when blue radiation is applied to the whole body or abdomen, it affects mostly to stimulate the sympathetic nervous system, but at the same time it also intensifies the irritability of the parasympathetic nervous system, and leads the vegetative nervous system into unstable condition." Its first action in this instance is to lower blood pressure, a condition that is followed after irradiation by a steady rise.

In the action of visible light on wounds, red light tended to accelerate healing and blue light to retard it. (This might suggest that products like iodine really ought to be blue in color. For a blue stain on the flesh would absorb the red rays of light—which are beneficial—and reject the blue.)

Regarding the growth of tumors, Dr. Mizutani writes, "As for the effects of visible light upon tumor tissue within or without the body, red rays affect to prevent the growth and blue rays affect, though slightly, to accelerate it."

Further Responses

Hoffman is of the opinion that the body has a radiation sense. The skin must contain cells that have close association with the nervous system and a definite reaction to radiant energy. Tonus reflex seems to be in two directions. Yellow-green is the neutral point. Toward orange and red there is an attraction to stimulus. Toward green and blue there is a withdrawal from it. Infrared and ultraviolet, both invisible, also will cause reflex actions, lending further evidence that the body will react to color without even seeing it.

However subtle this may be, it does hint of many physical reactions to the spectrum. Kurt Goldstein writes, "Under red and green lights, movements are carried on with a different speed, without subjectively experiencing the change in speed. Likewise, the estimates of traversed distances vary as to length; seen and felt distances, time intervals and weights, are judged differently under the influence of different colors."

Thus, under the influence of red, the average person will tend to overestimate time. He will also judge weights as being heavier. Under the influence of blue and green, time will be underestimated, weights will seem lighter. There may be some explanation for this in the fact that warm hues are exciting, and excitement seems to pack more experience into a given length of time. Conversely, cool colors are subduing, and time speeds on with less notice paid to it.

Human senses, as well are more alert under red than under blue or green light. Dr. Gilbert Brighouse of Occidental College, Los Angeles, has tested the muscular responses of several hundred college students. Reactions were 12 per cent quicker than normal under red light, while green light retarded the response.

Psychiatry

Back in 1875, a European doctor by the name of Ponza sought to deal with insanity by fitting hospital rooms with colored-glass windows, colored walls and furnishings. Red and blue were the hues principally used. Of red he wrote, "After passing three hours in a red room a man afflicted with taciturn delirium became gay and cheerful; on getting up the day after his entry into the room, another madman who had refused all food whatever asked for breakfast and ate with surprising avidity." As to blue, "A violent case who had to be kept in a strait jacket was shut in the room with the blue window; less than an hour afterwards he became calmer."

Color may not be a cure for insanity, and today the application of fever therapy and insulin shock are achieving far more potent results. However, the magic of color in its influence on human moods seems to inspire constant inquiry. Quotation is made from further work more recently conducted at the Worcester State Hospital in Massachusetts (1938).

"The experiments now in progress are conducted in a small ward with a nurse and attendant in charge. A regular program is followed with periods for work, play, meals, and rest. The patients spend their entire waking day in this ward, which is lighted by seven 100-watt lamps, daylight being excluded. All light passes through filters of the color being tested. These filters are similar to the gelatin filters used in theaters for color light effects. Careful notes are kept on the reactions of both the patients and attendants. Three colors have been tried and one is now in use.

"Magenta used with disturbed patients had a quieting effect for several weeks with a diminution of efficiency after the initial period. There was an associated stimulating effect which contributed somewhat to a feeling of tension. However, when magenta was replaced by white light after a month, the patients became quite excited.

"Blue had a striking and prolonged quieting effect. Patients and attendants commented on its soothing effect. This color was the most effective of those used.

"Yellow was used with depressed, melancholy patients and had a very slight stimulating effect. Red used with the same group of patients produced more stimulation than the yellow. The reaction of the depressed patients to color was short and less obvious than that of the disturbed patients.

"The work is still quite in its experimental stage, but we are rather hopeful that some of the colors may prove to be useful adjuncts on certain wards where disorders of the two extremes of mood are being treated."

(The author has been advised, however, that further research has since been set aside.)

Color and its value in psychiatry have been elaborately and skillfully investigated by Felix Deutsch (see *Folia Clinica Orientalis*, Vol. 1, Fasc. 3 and 4, 1937). Deutsch is a physician, and his work is very fascinating indeed. "Every action of light has in its influence physical as well as psychic components." By using color to help correct adverse mental states that bring about physical distress, he has been able to speed his patients back to recovery. His procedure has been a simple one—merely to expose the patient to a strongly hued environment (of his own choice), and then to let it exert its influence in arousing thoughts and moods favorable to recovery. Though the treatment may to a large measure be psychic, nonetheless the results are physical and tangible things are experienced by the entire vascular system of the body.

Here, for example, are two case histories. One patient troubled with anginal fear complained of shortness of breath, air hunger, and palpitation of the heart. She feared the return of a spasm which years before had caused her to lose consciousness. An examination of her heart revealed a fairly normal condition. She had a slight thyroid enlargement. Her pulse rate at the time of her examination was 112, her blood pressure 115/70.

This patient was placed in a red environment. (Green as a color was decidedly unpleasant to her.)

First session: pulse 112; after treatment pulse was 80.

Second session: pulse 92; after treatment pulse was 76.

Third session: pulse 92; after treatment pulse was 80.

Fourth session: pulse 84; after treatment pulse was 74.

During following sessions her pulse was always 74. The patient experienced a comforting sensation of warmth. Her insomnia disappeared and she felt restored calmness.

In a second case, the patient complained of attacks of weakness, shortness of breath, and pressure sensations over the chest, which led to fear of choking. When she was admitted, her blood pressure was 245/125. Medical therapy did not produce any changes in this reading, nor did it lessen the patient's subjective complaints.

When she was placed in a green room for short periods, however, the following reduction in blood pressure took place.

First session: 250/130; later 210/125.

Second session: 245/130; later 205/120.

Third session: 240/125; later 205/120.

Fourth session: 220/120; later 195/110.

Fifth session: 210/115; later 210/110.

Sixth session: 200/110; later 180.

Seventh session: 195; later 180.

"In the course of other sessions the blood pressure fell to 180 mm. Hg and was attended by subjective, relative feelings of well-being."

Deutsch assumes that biological actions take place following the use of color. He summarizes his conclusions in these four points:

1. Color brings about a reflex action upon the vascular system, if only through the feelings and emotions.

2. The effect achieved is not specific for any one or any certain hues. Warm colors may calm one person and excite another. Cool colors may likewise be stimulating to one person and passive to another.

3. Irradiation with red or green light may produce an elevation of blood pressure and a quickening of pulse rate. Or the opposite may take place, depending on the particular psychic make-up of the individual.

4. "An organic, nonoptical color sense has not been proved so far." However, the response that follows exposure to color may be decidedly organic in effect. "The psychic process which is brought into play here is easily stated: the colored light changes the environment. Through the changed appearance of the environment the individual is lifted out of reality." He is on the road to recovery.

Warmth and Coolness in Color

In looking at the spectrum, the average person will see its colors as warm or cool. This psychological response is to be observed frequently in the reaction of people to an environment. Murray and Spencer, in their book *Colour in Theory and Practice*, report such an instance. "A room was built as a rest room for the employees of a certain firm, and was painted inside in slate blue and gray. Although the air was conditioned and maintained at a fixed temperature, the employees complained that the room was cold. On advice, the room was repainted in warmer colors, brown and orange. On the room being put into use again, the employees reported that the alternations had produced conditions that were comfortably warm. Nevertheless, the temperature, which was maintained the same as that of a large number of other rooms in the same building, had not been altered."

Some investigators have attempted to set up laboratory conditions to measure this reaction to warmth and coolness in terms of physiological and therefore measurable responses. One such experiment was made some time ago without success. This is not at all surprising. The mere fact that most of us think of colors as being warm or cool is evidence enough. I suppose that you and I would not feel hungry if we were put before a plate of meat while various instruments were clasped to our wrists or slipped under our tongues to measure saliva flow or the quickening pulse of appetite.

S. M. Newhall of Johns Hopkins remarks, "Unless the test situation is sufficiently similar to the actual situation, the test results can have no practical value in application to the actual situation." Newhall in studying 297 observers found that the warmest hue was red-orange, this reaction being very marked. In settling upon the coolest of hues, however, the general region of blue and green was chosen. "The relatively great hue range covered by such perceptually cool objects provides opportunity for the psychological association of coolness with a relatively great range of hues."

Synesthesia

There are many persons who associate colors quite spontaneously with other things. They may think of the letters of the alphabet as being hued. Numbers may suggest colors and so, too, many sounds, musical notes, the days of the week, and a wide variety of other tokens. This faculty, which seems to be inherent in the psychic make-up of an individual and to exist without change during his lifetime, is called synesthesia.

Over fifty years ago, Francis Galton made a study of these "color-thinkers." One man associated colors with numerals as follows: 1 was black, 2 yellow, 3 a pale brick red, 4 brown, 5 dark gray, 6 reddish brown, 8 blue, 9 brown. A woman "saw" the letter *A* as white, *E* as red, *I* as yellow, *V* as purple, and *Y* a dingy hue.

With music, synesthesia often accounts for many other strange notions. In a program for the Boston Symphony Orchestra, some years ago, Philip Hale commented on a few of the "color-hearing" reactions of musicians. He told that Raff held the tone of the flute to be intensely sky blue. The oboe was clear yellow, the trumpet scarlet, the flageolet deep gray. The trombone was purplish to brownish; the horn, greenish to brownish; the bassoon, a grayish black. He remarked that A major was green to one musician and that another felt the hue of the flute to be red rather than blue, as it was to Raff.

In 1890 a woman was found to whom the music of Mozart was blue, that of Chopin yellow, and that of Wagner a luminous atmosphere with changing colors. To another subject *Aïda* and *Tannhauser* were blue, while *The Flying Dutchman* was a misty green.

Liszt is credited with a number of pet phrases, which perhaps had their origin in synesthesia. "More pink here, if you please." "That is too black." "I want it all azure." Beethoven is said to have called B minor the black key. Schubert likened E minor "unto a maiden robed in white with a rose-red bow on her breast." One Russian composer said, "Rimsky-Korsakoff and many of us in Russia have felt the connection between colors and sonorities. Surely, for everybody sunlight is C major and cold colors are minors. And $F^{\sharp}$ is decidedly strawberry red!"

Eidetic Imagery

Finally, there is another singular human talent, a bit rare, which the psychologist terms eidetic imagery. E. R. Jaensch, who has written a book on the subject, states: "Eidetic images are phenomena that take up an intermediate position between sensations and images. Like ordinary physiological after-images they are always *seen* in the literal sense. They have this property of necessity under all conditions and share it with sensations."

Eidetic imagery is the gift of childhood and youth. While seemingly akin to the supernatural, it is nonetheless a sensory reality. The child playing with his toys may be able to project living pictures of them in his mind. These may not be mere products of the imagination. They may be far more tangible, with dimension, color, movement. They are "lantern slides" of the eye and brain, projected into definite, localized space. They are images as real as projected lantern slides.

What occurs? A person endowed with the faculty may glance for a few seconds at a showing of pictures, words, letters, or colors, and then be able to hold the image clearly before him. H. Klüver writes: "For example, an eidetic child may, without special effort, reproduce symbols taken from the Phoenician alphabet, Hebrew words, etc. Or a person with a strong eidetic imagery may look at a number of printed words for a while and then go to the dark room and revive the text eidetically. It is possible to photograph the eye movements occurring during the reading of the eidetic text."

Because the phenomenon seems to vanish with age, being likely to disappear during adolescence, the adult mind, capable of dealing with it, relegates it to the fervid period of childhood. Nevertheless, images are seen. Pictures stand before the eyes and details are distinguished in them that may be counted and identified in hue.

We may seem to be wandering afield from the problems of selling color; yet, the more one learns of people and their reactions to the world of color, the broader his viewpoint should be and the closer to an understanding of those forces, curious or otherwise, that influence people.

We are all sensitive to color. We take it seriously, praising its goodness and condemning its evil; and at times it rouses us to very strange actions. The following item recently appeared in the *Long Island Star-Journal.*

"Camden, N. J.—City Commissioners were pondering an amazing complaint which was handed them today by residents of Newton Avenue.

"The complaint was registered against the newly painted two-story house of one of their neighbors. Residents said the color of the house was affecting their health and petitioned the commission to make the owner change its color. Said the request:

"'The building is a bright, glaring yellow—with red and black lettering—which reflects a ghastly, sickening glare of yellowish light in the front of the houses opposite. It creates a nuisance which endangers health, physical and mental well-being.'"

Chapter 14

THE ROMANCE OF COLOR

THIS is a chapter on symbolism, tradition, romance, and personality. The subject of color seems to have almost endless ramifications and to touch upon life in almost every quarter, for color is rich in lore, rich in meaning and purpose. If the present chapter seems to have loose threads, kindly regard it as an attempt to weave together a wide diversity of interests, related and unrelated, all fitting into some part of the warp and weft of color.

The Senses

The appeal of color calls forth many associations with the other senses besides sight. Notes have already been presented on taste and appetite. The eye is highly discriminating about the appearance of foodstuffs destined for the stomach of the body. Butter, for example, must have a pale, creamy tint. Let it be too yellowish, and it will look old and rancid. Let it be too white, and it will "savor" of oleomargarine. Let it be too lemonish in tinge, and it will look odd and distasteful. A succulent orange is distinguished for its rich hue. So strong is this endowment to the average mind that fruit grown for table consumption in one part of the country is dyed with pigment to resemble the naturally fuller chroma of fruit grown in another part of the country!

The observation has been made that "We would resist blue mashed potatoes, purple bread, and yellow-green steak, no matter how delicious." Why not, in the merchandising of foods, in packages, advertising, displays, respect this psychological trait in human nature?

As to the sense of touch, colors may appear warm or cool, dry or wet. What better to bespeak the pleasures of an air conditioner than blues and greens, or the comfort of a heating system than red, orange, and yellow?

In the sense of smell, the sweetness of perfume is surely lavender, pink, or delicate yellow. Here, the odor of June may recall green; or blue may seem most refreshing in an after-shave lotion. I wonder, however, who made the mistake of a brown cake of toilet soap noticed one day on the bargain counter of a retail store?

As to the sense of hearing, the phenomenon of synesthesia has already been mentioned. According to the extensive research of T. F. Karwoski and H. S. Odbert, the majority of human beings will associate slow music with blue, fast music with red, high notes with light colors, deep notes with dark colors. More subtle, "The horizontal dimensions might be related to the development of music in time; the vertical dimension to changes in pitch. A third dimension of depth may eventually be available to denote volume or intensity." In other words, music moves along quickly or slowly depending on its tempo. It jumps up into tints for high notes or drops down into shades for low notes. When it is *fortissimo*, the colors are near, heavy, and bulky. When it is *pianissimo*, the colors are filmy and far away.

Colors also suggest definite forms. Although these associations may not be so fixed in most minds, they are perhaps worthy of mention here for those who may feel (as the author does) that shapes and hues bear resemblance to each other:

Red suggests the form of the square and the cube. It is hot, dry, and opaque in quality. It is solid and substantial. It is extremely advancing and holds the strongest of all attractions to stimulus. Because it is sharply focused by the eye, it lends itself to structural planes and sharp angles.

Orange suggests the form of the rectangle. It is less earthly in quality than red, more tinged with a feeling of incandescence. It is warm, dry, compelling. Optically, it produces a sharp image and therefore lends itself to angles and to well-defined ornament.

Yellow suggests the form of the triangle or the pyramid, with its apex down. It is the color of highest visibility in the spectrum and therefore sharp, angular, and crisp in quality; but it is more celestial than worldly. It is without solid base, lofty, and reaches upward into space.

Green suggests the form of the hexagon or the icosahedron. It is cool, fresh, soft. It is not sharply focused and therefore does not lend itself to much angularity. It is a *big* color and can dominate the eye without distressing it.

Blue suggests the form of the circle or the sphere. It is cold, wet, transparent, atmospheric. Blue is retiring, dignified, and tends to create a blurred image in the eye. While it may have bulk, it does not seem to lend itself to angularity.

Purple suggests the form of the oval. It is even more refined than blue. The eye finds difficulty in focusing an image of it sharply. It is thus soft, flowing, and never angular. Unlike blue, however, it is not so lofty but clings closer to earth.

Other Associations

Language, being an expression of human thought and emotion, becomes significant when studied for its slang, metaphors, similes, and colloquialisms. For a quick review of color jargon, consider these examples.

At once, red is the passionate and ardent hue of the spectrum, marking the saint and the sinner, patriotism and anarchy, love and hatred, compassion and war. The word red in Russian means beautiful. Man must paint the town red. He sees red and, thereupon, works himself into a purple rage. The red lamp is a brothel, the scarlet woman a prostitute. Red-hot news is in his daily paper. He reads of red tape and the drawing of red herrings across the line. There are reds in Russia, red-letter days, businesses in the red, and bums without a red cent to their names.

Yellow, man despises for the most part. The scoundrel is yellow. Sensational journalism is yellow—an expression that sprang up in 1895, when the *New York World* ran a picture in yellow as an experiment in printing.

Green is the demon of jealousy. Greenbacks are paper currency. Greeners are inexperienced workers; greenhorns are dolts from the country.

Blue is indispensable. The expression to "feel blue" is a masterpiece of vivid statement. There are blue laws, blue gloom (for reformers), blue Mondays, bolts from the blue, blue bloods, and bluestockings. You can be true blue, curse the air blue, or experience something once in a blue moon.

As to other hues, you can have a purple time and, if you are not careful, be done up brown. You can cast black looks or feel black despair. There is blackballing and blackmailing. There are blackguards, blacklegs, black coats (clergymen), black boxes, and black books. To say that a man is white all the way through—or just white—is an Americanism dating

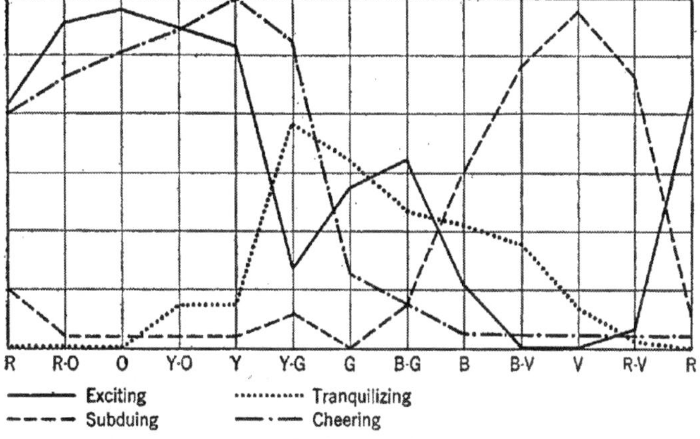

The affective value of colors. Red-orange is seen as the most exciting of hues, violet as the most subduing, yellow-green as the most tranquil, and yellow as the most cheerful. (*After Wells.*)

back to 1877. There is also the white man's burden, the white-haired boy, and the white feather—each, telling enough in its implication.

A number of psychologists have carried out research to determine the moods that people associate with the colors of the spectrum. Few of them, however, seem to be aware of the fact that a color may have contradictory qualities, depending on the particular viewpoint of the observer. Green is an excellent case in point. As seen objectively, it is cool, fresh, clear, and altogether pleasing. But green illumination shining on the human flesh may inspire a subjective viewpoint that will instantly make the color repulsive. Thus, no list of color associations is adequate unless it takes into consideration these subjective as well as objective aspects; for reactions will differ as a person associates color with the outside world or with himself.

While warm colors are not greatly different objectively and subjectively, cool hues are almost antithetical from the two viewpoints. Red, however, may seem far more exciting as applied to oneself than to external objects. Blues and greens, which appear peaceful in one aspect, may be terrifying in another.

Thus the moods conveyed by a single color may be diverse. This modern emotional symbolism is presented in a separate tabulation. Here the major colors are described in their general appearance, their mental associations, direct associations, objective impressions, and subjective impressions. The author is hopeful that most of the associations will agree with the personal experience of the reader.

Symbolism

The complete story of color symbolism would fill a pretty thick book. To be thorough, it would have to embrace practically all the civilization, religion, art, science, and custom known to mankind since the beginning of time. While such an exposition might have literary interest, the endeavor would be of small practical benefit, inasmuch as the boots of time have trampled the vast majority of these curiosities and amenities into the dust of history.

To give the reader a taste, however, a few survivals are listed. All, of course, has not been lost. The policeman probably wears a blue uniform because of a tradition established when the Romans dressed their servants in that hue. Black for mourning, red as a universal symbol of danger, yellow for quarantine—these are old memories. The hues of flags, military uniforms, religious vestments, racial and national costumes, holiday festivals, mostly trace back to a meaningful and devout symbolism.

The Roman Catholic Church, for example, recognizes five canonical hues. "By a law of her liturgy the Church directs that the vestments worn by her sacred ministers, and the drapery used in the decoration of the altar should correspond in color to that prescribed for the Offices of the day."

White, the symbol of light, signifies innocence and purity, joy and glory. Red, the symbol of fire and blood, signifies

The Modern Symbolism of Color

Color	General appearance	Mental associations	Direct associations	Objective impressions	Subjective impressions
Red	Brilliant, intense, opaque, dry	Hot, fire, heat, blood	Danger, Christmas, Fourth of July, St. Valentine's, Mother's Day, Flag	Passionate, exciting, fervid, active	Intensity, rage, rapacity, fierceness
Orange	Bright, luminous, glowing	Warm, metallic, autumnal	Halloween, Thanks-giving	Jovial, lively, energetic, forceful	Hilarity, exuber-ance, satiety
Yellow	Sunny, in-candescent, radiant	Sunlight	Caution	Cheerful, inspiring, vital, celestial	High spirit, health
Green	Clear, moist	Cool, nature, water	Clear, St. Patrick's Day	Quieting, refreshing, peaceful, nascent	Ghastliness, disease, terror, guilt
Blue	Transparent, wet	Cold, sky, water, ice	Service, Flag	Subduing, melan-choly, con-templative, sober	Gloom, fearful-ness, furtive-ness
Purple	Deep, soft, atmos-pheric	Cool, mist, darkness, shadow	Mourning, Easter	Dignified, pompous, mournful, mystic	Loneliness, despera-tion
White	Spatial—light	Cool, snow	Cleanliness, Mother's Day, Flag	Pure, clean, frank, youthful	Brightness of spirit, normality
Black	Spatial—darkness	Neutral, night, emptiness	Mourning	Funereal, ominous, deadly, depressing	Negation of spirit, death

From *The Story of Color.*

charity and generous sacrifice. Green, the symbol of nature, signifies the hope of eternal life. Purple, the gloomy cast of the mortified, represents affliction and melancholy. Black is symbolic of the sorrow of death and the somberness of the tomb.

In Freemasonry there are glints of colors once sacred to Egypt, Babylon, Judea, and the teachings of the Cabala. Craft Masonry is blue, the symbol of eternity, beneficence, and charity. Royal Arch Masonry is red. In the Scottish Rite eight color groups are recognized: black, white, blue, red, green, yellow, blue and yellow, pink and blue. Strict fundamentalism is not observed, unfortunately. One Masonic en-

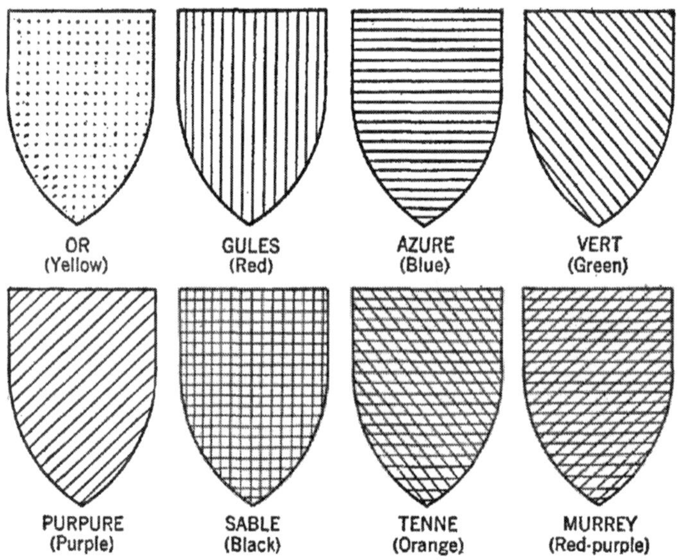

The tinctures and "hatchings" of heraldry. Black and white symbols used in the designation of colors.

cyclopedia states: "This scheme is motley, for the draperies of Lodges and Chapters may or may not correspond with the insignia of Officers and Members."

Another interesting survival is that of heraldry—still a part of culture in England. Britain has its coats of arms which have been extended even to corporations such as the British Broadcasting Company, Lloyds of London, the Worshipful Company of Haberdashers, and to municipalities, schools, Rugby teams.

Nine colors, or *tinctures*, are recognized, and the symbolism is as follows:

Gold (or yellow) is called *or* and stands for honor and loyalty. Silver (or white) is called *argent* and represents faith and purity. Red is called *gules* and is a token of courage and zeal. Blue is called *azure* and signifies piety and sincerity. Black is called *sable* and means grief and penitence. Green is called *vert* and means youth and hope. Purple is called *purpure* and means royalty and rank. *Tenné* is an orange color signifying strength and endurance. *Murrey*, or *sanguine*, is a reddish purple and represents sacrifice.

Heraldry, incidentally, has left an efficient means of identifying colors in black and white. The various *hatchings* used to signify hues in drawings and engravings are separately printed. These designations in black and white have many conveniences outside the field of heraldry and are frequently used to indicate colors where actual reproduction of them is impractical.

In modern times, color is made to serve a few more humble and honorary functions. The hues of the music scale, as conceived by Newton and still recognized, are red for note C, orange for D, yellow for E, green for F, blue for G, indigo for A, and violet for B.

In Brazil, a country abounding in precious and semiprecious stones, professional rank is denoted by gems. The physician wears an emerald ring, the engineer a sapphire, the lawyer a ruby. The professor affects the green tourmaline, the dentist the topaz, the commercial man the pink tourmaline.

In America (since 1893) the universities and colleges have recognized a code of color to identify their major faculties. These hues make up part of the insignia of the learned and are worn in gown, cape, braid, or tassel as follows: scarlet represents theology; blue is for philosophy; white is for arts and letters; green is for medicine; purple is for law; golden yellow is for science; orange is for engineering; pink is for music.

In stock, horse, dog and cat shows, the first prize award is generally a blue ribbon; the second prize is a red ribbon; the third yellow or gold; the fourth white. Purple may be used for winner over all classes, and green for special prizes.

As to the colors of the months and the seasons, the following list is submitted:

Spring: Pink and green
Summer: Yellow and blue
Fall: Orange and brown
Winter: Red and black
January Black or white
February: Deep blue
March: Gray or silver
April: Yellow
May. Lavender (lilac)
June: Pink (rose)
July: Sky blue
August: Deep green
September: Orange or gold
October: Brown
November: Purple
December: Red

Color traditions recognized during holidays are quoted from the sales records of a manufacturer of holiday novelties

New Year's

Color	Per Cent	Color	Per Cent
Red	40	Green	14
Blue	34	All others	12

St. Valentine's Day

Color	Per Cent	Color	Per Cent
Red	73	Gold	11
Blue	7	All others	9

St. Patrick's Day

Color	Per Cent	Color	Per Cent
Green	89	White	9
All others			2

Easter

Color	Per Cent	Color	Per Cent
Yellow	34	Orchid	14
Light green	31	Cerise	11
All others			10

Halloween

Color	Per Cent	Color	Per Cent
Orange	68	Green	4
Black	17	Yellow	3
All others			8

Christmas

Color	Per Cent	Color	Per Cent
Red	52	Blue	8
Green	31	All others	9

Symbolism often has importance in selling color. Beyond the fact that it offers a chance for romance, it may at times actually cause trouble. There are three stories to tell about China. In one instance, the white service stations of an American oil company had to be changed in hue because of a strong national association with white as a token of mourning. Nor could pins in blue packages be sold well among the Chinese for the same reason. The famous Flying Red Horse, a brilliant red, had to be changed from a mare to a stallion—because of the high regard held for masculine "principles" among the Chinese, and for red, which is a masculine color.

Colors and Period Styles

For a review of color traditions in home furnishings and wearing apparel the reader is referred to two excellent books by Elizabeth Burris-Meyer, *Historical Color Guide* and *This Is Fashion*. Both volumes are written in scholarly fashion and are generously illustrated with actual color chips.

Practically all period styles draw from two wellsprings—the classical and rather severe traditions of ancient Greece and Rome, and the more sumptuous and baroque elegance of the Renaissance.

The French periods, notably that of Louis XV, are luxurious with brilliant and refined hues—light blues, violets, odd greens, purples, grays, gold—probably through the influence of Madame de Pompadour, whose favorite color was rose. The Louis XVI style goes back to simpler classical ideals, as to both design and color. Under the influence of Marie Antoinette, colors are delicate and affected. Under the influence of the eminent painter David, there is extreme severity and formality. French Empire was, to a large extent, inspired by Napoleon's visits to Italy. Napoleon showed preference for red, green, white, and gold; Josephine, for the more delicate tints of blue, violet, purple, tan.

The English periods are more temperate. The Georgian style is largely Roman and Pompeian, although more reserved. The Adam style is directly Pompeian. Josiah Wedgwood's classical Greek ceramics are simple in color, severe, and pure in tradition. Adam, however, introduced less brilliant colors

—soft blue, pale yellow, lilac, delicate gray, blue-green, yellow-green, and pink.

The craftsmanship of Chippendale, Hepplewhite, and Sheraton also emphasized the subtle beauty of wood tones, which became definitely tied in with interior schemes. The Victorian style is flamboyant. Colors were generally of low key—brown, purplish red, deep green. The wide use of purples and magentas has distinguished the era as the Mauve Decade.

In America, the colonial style had much individuality. The most common hue in the seventeenth and early eighteenth centuries was a dark red, called Spanish brown. This was made from a pigment dug out of the earth. It was nearly always used for the priming coat on the exterior of a house and was often the sole paint applied. In 1769, John Gore of Boston advertised "Very good red, black, yellow paints, the produce and manufacture of North America."

For interiors, blues and greens were favorites, often with the green made olive or the blue grayed by the addition of black. Vivid colors were found on spattered floors, plastered walls, and woodwork. White, gray, pearly tones, and stone shades seem to have been favored for meetinghouses.

The Williamsburg style, very beautifully preserved and recorded in Mr. Rockefeller's distinguished restoration, has colors of a soft, deep beauty. In predominance are grayish and olive greens, warm tans, browns, with some dark, rich reds and greens. The general feeling is on the subdued side, with an occasional use of off-whites. White, in fact, is today perhaps overemphasized in colonial revivals, although it did have a considerable vogue in the late eighteenth century. Mostly, the colonists preferred deeper and stronger colors. These they applied with a simplicity and candor that well fitted their architecture.

In the matter of early American dress, Elizabeth Burris-Meyer mentions a few interesting restrictions and customs. In 1634, Massachusetts had a law forbidding gold and silver stitching and accessories. Town records were full of penalties for the wearing of ribbons, broad-brimmed hats, and lace. Among the Mennonites of Pennsylvania, red was worn by young girls, rose by unmarried women, blue by the married, and white by widows. "An unmarried woman might wear all white with rose ribbons on her close-fitting cap. The men usually wore brown or gray."

Colors and Personality

Much has been written about the harmony of color in dress. Yet, because such advice so frequently gets lost in clouds of vain rhapsody, a brief and humble statement may be refreshing. There are sure and scientific ways of relating colors to complexion tones and to the hues of eyes and hair. Although glamour may be the end in view, the means to such an end may be quite sober and deliberate.

The all-American brunette (so-called) is perhaps the most prevalent and typical of national types. Her hair is brownish in quality, neither light nor dark. Her complexion is creamy, with good natural tinting that is neither bleached nor olive in cast. Her eyes may be brownish or hazel, or perhaps on the grayish blue side. None of her coloring is extreme; nor should the colors she wears be extreme, except in small touches. Her one best hue is a soft blue-green. Other suitable tones are natural beige on the tan side, maroon, navy, soft rose, periwinkle blue, grayish violets, and dull flesh tones. Accent colors may well be bright, preferably to match the glint of her eyes.

The composite blonde has hair of a fairly deep golden cast, frequently streaked, like molasses taffy. Her skin is usually creamy or toned with a blush of tan. Her eyes are bluish, greenish, or grayish—and also composite, in that they are seldom extreme in hue. She must be careful of rich colors, which, by contrast, will compromise her own endowments. Tints are dangerous, shades more dramatic. Her best color is a soft periwinkle blue. Other good hues are white, soft grayish greens, wine, grayish blue, navy, olive green, purple, and black.

The vivid brunette, with fair skin and deep brown eyes and hair, has almost the whole world of color to command. She may like red best of all, but will probably look best in green, blue, or violet. Hers is about the only type that can wear yellow and orange effectively. Grayish tones also become her, especially when set off by rich, brilliant accents and accessories.

The vivid blonde, who has fair skin, fair hair, and blue or brown eyes, must be judicious with tints. Gray is spectacular, black always startling. Tones generally on the greenish, bluish, or violet side of the spectrum—grayish or deep—are mostly appropriate.

The redheaded titian is best in green.

Those who have gray hair are almost always seen at their best in blue.

Character Analysis through Color

In a closing glance at romance, a few character readings reprinted from a booklet prepared by the author for the Paint Merchandising Council of Chicago are given below. Though the idea is somewhat extravagant, it has some basis in fact, being a summary of observations gathered over the years. It is included here chiefly for entertainment, although various commercial applications may suggest themselves to the reader. A more elaborate "treatise" will be found in *Character Analysis through Color*, a book written by the author under the pseudonym of Martin Lang.

RED: Here is the vital color. Love red in a big and generous way and you have real courage before life. You are vigorous and given to action, for life means much to you. Do you like sports? Are you inclined to be very positive in what you have to say?

In affairs of the heart you are rather likely to be fickle. Because of your eagerness for excitement you may break many old hearts in your glamorous efforts to win new ones. But people who understand you recognize that you are a very human person, after all, and will more readily forgive your faults than you will forgive theirs.

You are quick to judge people. First impressions count with you and very often cause you to act emotionally rather than mentally.

MAROON: If you like maroon better than red, the chances are that you are a vital person at heart, but have been tempered somehow by strict training, severe parents, an exacting wife or husband, or by a series of hard trials in life. Being something of a red type, you yet are steadier and not so reckless with your friends or with the world in general.

It is rather certain that you are a really great person, that you have a compassionate heart, but that you may, in perfect innocence, demand considerable from others. People are sure to like you—not all people, of course, but most. In your turn, you are likely to seek many friends and perhaps arouse some jealousy on the part of those who know you intimately; but you make an excellent husband or wife. You have a pleasant way of confessing your sins and then sweetly starting all over again to commit more.

PINK: The gentler way of life is the one for you. Surely, if you like pink, you are a woman and have much feminine charm. Were you very smart as a child? Were you the idol of doting parents or relatives? Did those who loved you keep you in pretty, starched dresses and hold from you all the sordid details of a nasty world?

You have good taste in friends and clothes and a very gracious social manner. You like people who are cultured and refined. While you may have sympathy for persons in circumstances less fortunate than yours, it is not your habit or desire to associate with anyone outside your own genteel class.

You adore affection but sometimes are a bit frightened by it. People are drawn to your charms, and sometimes you have rather a hard time holding them off.

ORANGE: Choose orange and you are a person to be envied. For you are a "hail fellow well met." Perhaps not so ardent and so passionate as the red type, you still have a grand love for life.

Your taste runs to cheerful friends and good foods. You are just as much at ease among sinners as among saints. In fact, much of your life is devoted to things social. Yet, you should be careful. People may say behind your back that your last friend is always your best friend.

You are not one to be alone. You like eminent people, whether they are prizefighters or presidents. You want life to surround you, warm and mellow, like the orange glow from a fireplace.

You make the ideal bachelor or spinster, perhaps because you are inclined to know a little about many people and not very much about one or two.

YELLOW: This hue is often picked as a favorite color by very intelligent people. If you like yellow, you are probably high-minded and, perhaps, a person who is drawn into reforms or strange cults.

You have a beautifully controlled temper. Yet the temper is there, nonetheless, and you are quite capable of putting people in their places if you feel so moved. People you meet for the first time may not always think you the most friendly soul in the world; but those who know you intimately will respect and cherish your profound character.

You long secretly for the admiration of others. While you outwardly resist flattery, you inwardly glory in it. You are inclined to live alone mentally. You are a true friend and a rare soul who can keep secrets.

GREEN: This is nature's color, preferred by human beings who are likewise fresh and natural in personality. If you like green, you are sure to have a rather broad interest in the world, to be aware of the problems of life, to be tolerant and somewhat liberal.

Because your mind is clear and full of varied attentions, you are an agreeable person. You like to play bridge, to shop, to read all the best selling books. You respect money and are anxious to improve your standard of living. In this, perhaps, you are like many others. Yet the chief difference in you is that you delight in friends, travel, sensible luxuries, without being either miserly or inclined to put on a false front. You are good and normal, and relish the scandals of others while carefully avoiding them within your own household.

BLUE-GREEN: A few persons will choose blue-green as a favorite hue. If you are one among them, you no doubt are a rather fussy person. This color indicates a well-ordered life, a sensitive nature, and a person who has pretty set notions about many things.

You are quite capable of managing your own affairs without asking for advice or help. You are orderly and want things just so. You can be generous without bothering yourself about rewards. Yet you are emotionally a trifle cold and not the easiest person in the world to get along with.

Being your own master, perhaps now and then you commit a few sins that may seem pretty bad to others, although to you they are quite harmless. Are you good-looking? Surely you have excellent taste in clothes. You are sweet and charming in your manner (when you so desire) and wholly self-possessed.

BLUE: Here is the color of conservatism and dignity. You are sensitive to others, sensitive to yourself. You have real weight to your character and never enter into any silly enthusiasms without careful thought.

You are not one to monopolize conversations, although you can hold your own when you once make up your mind to do so. You are cautious in word, dress, and action. What troubles you most, however, is an inability to let go in moments of excitement. Perhaps situations and personalities bother you unduly.

When you sin, your conscience is bothered, but you *will* sin just the same. This is because you have a rational mind and know how to justify yourself.

PURPLE: Two different types of people like purple—very profound beings and those who merely wish to appear so.

Choose purple and you are, no doubt, a mystery to yourself as well as to others. Aristocratic and artistic people often favor the hue for its exclusive dignity. Such persons are generally satisfied with themselves, sometimes conceited, and nearly always capable of great things.

If you like purple, you no doubt lay claim (if only in your heart) to a rather superior attitude toward the world; but, if you are extremely clever, are you also inclined to be a trifle lazy?

Purple types are easy to live with. They are neither too bold, like red, nor too strait-laced, like blue. Yet they must beware of any "lavender-and-old-lace" tendencies, which may snare them if they don't watch out.

BROWN: Choose brown, and you are of the earth, substantial, dependable, steady. You avoid things showy or gaudy. There is always an admirable sameness about you, which may lead people to say, "Ah, you never change a bit over the years!"

You are not one to take the spotlight. Your brain is slow but sure in its action. You never shirk responsibilities. People could lend you money and never have to worry. Discipline is no hardship. You can take it as well as mete it out.

In sports, you are more inclined to sit in the grandstand than to appear on the playing field. You like good, wholesome foods; well-tailored clothes; reliable and trustworthy friends. You do not seem to be very curious. New things make only a moderate appeal to you.

Chapter 15

THE SPECIFICATION OF COLOR

MOST books on color start with an exposition on the physical nature of light and radiant energy. This book ends with such a discussion. There is good reason for this. Color as sensation, as we explore its mysteries in human consciousness, has little to do with the science of physics. Deane B. Judd writes, "Color is not a stain, a specimen, or a spectrophotometric curve." Nor is it a wave length or a shower of electrons upon the retina.

The introduction to a cookbook would hardly be relevant if it were devoted to a technical study of the chemical composition of meats and vegetables. To develop a good color scheme—or to bake a good cake—demands first attention to the strange qualities of human appetite.

However, a few notes on matters of physics will probably interest the reader, as well as serve for a prologue to the subject of this particular chapter.

The Physical Nature of Color

The exact nature of light and color is still an enigma. LeGrand H. Hardy writes, "A wave theory, an electrical theory of matter, or an atomic theory of energy can be made perfectly legitimate and defensible." Newton, who formulated the first plausible explanation, supposed that light was generated by an emission of particles. Though he was opposed by Hooke, who championed a wave theory, his ideas prevailed for a full century and a half. Then the wave theory again came into vogue, to be still later questioned by exponents of a corpuscular theory. In the midst of all this wavering, Sir William Bragg was once led to remark that scientists ought to use the wave theory on Mondays, Wednesdays, and Fridays, and the corpuscular theory on Tuesdays, Thursdays, and Saturdays.

Through the efforts of such men as Planck and Bohr, however, scientists today are in pretty fair agreement. Radiant energy is said to be propagated through space in the form of electromagnetic waves. The visible portion of this energy is seen as light. A substance excited to luminosity radiates energy in the form of electromagnetic waves. This collection of waves is characteristic of the substance and may be analyzed in a spectroscope. In addition, the waves that a substance radiates when excited are identical with those it will absorb when radiant energy falls upon it!

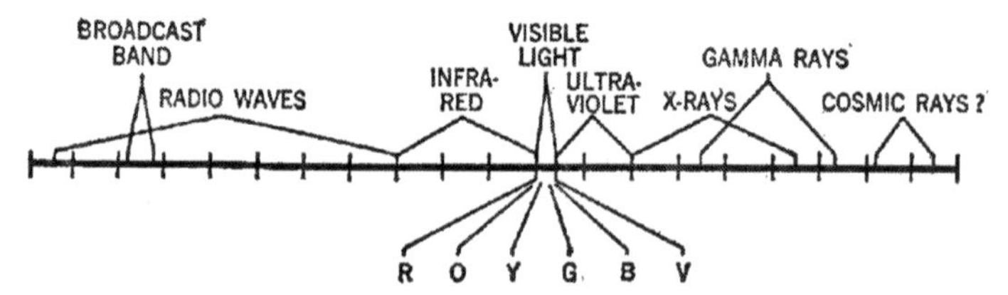

The electromagnetic spectrum. The visible portion occupies one octave out of seventy.

Radiant energy, however, has not only a wave structure but a corpuscular structure, as well. This means that radiant energy has tangible substance. It has mass and may be "bent," for example, by the force of gravity. In a sense, one is justified in speaking of a pound of light energy, just as he might speak of a pound of sugar. The sun emits rays the material pressure of which is said to be equal to about 250 tons a minute. Feeble though their strength may be, these rays can push things.

The complete spectrum of electromagnetic waves is charted in an accompanying diagram. The visible waves, which we see as light, occupy but a narrow section, about one "octave" out of 60 or 70. Waves of greater length than the visible waves correspond to infrared and radio rays. Waves of lesser length are found in ultraviolet, X-rays, radium rays, and cosmic rays.

All this energy travels at the same rate of speed—186,000 miles a second—and differs only in length of wave as measured from crest to crest. Radio waves measure about 2,000 feet from crest to crest. Waves of red light (longest of the visible spectrum) measure about 1/33,000 inch. Waves of violet light (shortest of the visible spectrum) measure about 1/67,000 inch. Fabulous though these figures may seem, measurements of the speed of light and of wave lengths are extremely accurate and are accepted throughout the scientific world.

There is, however, one joker in the deck—the ether. Radiant energy, producing light and color, to have substance and a wave length must travel in or upon something. Although science assumes the existence of an ether, men like Michelson and Morley have failed to detect it. Light will move at the same speed past a moving observer, regardless of whether he is going with it or against it. Sir James Jeans writes, "The ethers and their undulations, the waves which form the universe, are in all probability fictitious."

It is much as if physics measured the size, shape, and speed of waves set up by a ship, and then was unable to prove the existence of the sea upon which the ship sailed.

Color Vision and Color Blindness

LeGrand H. Hardy writes, "What happens when light acts upon a photoreceptor is in essentials unknown." There are many theories of color vision, some mechanical, some electrical, some chemical. There are, for example, two types of nerve endings on the retina of the human eye—rods and cones. The rods, distributed mainly on the outer boundaries, give rise mainly to sensations of brightness. The cones, concentrated in the center of the eye (fovea), give rise to sensations of hue.

When light strikes the eye, various things take place. Electrical currents are produced in the optic nerve. Pigment granules in the retina may "migrate," the pigments hiding when darkness exists and coming up into the spaces between the rods and cones under the stimulation of light. Again, the cones themselves may contract under the influence of light, while the rods may swell. The visual purple, which floods the retina, will grow more acid under the stimulation of light, and it will also bleach out.

Color blindness is usually accounted for as a visual shortcoming, the cones of the eye for some reason being inactive to certain radiant energy. John Dalton, an English chemist, was the first man to give a lengthy description of it. During his early years, he had attributed his errors in judgment to an ignorance of color terms. He discovered later that he saw no difference in color between a laurel leaf and a stick of red sealing wax. He compared the color of a scarlet gown to that of the trees and once attended a Quaker meeting in a drab coat and flaming red stockings. So famous was his report, that Daltonism became the common synonym for color blindness.

The affliction is of several types. Total color blindness is exceedingly rare, and is usually attended by a general deficiency in the perception of light and form. Inherited red-green blindness, the most prevalent type, is seldom accompanied by any other abnormalities. Persons who have it see detail and brightness with normal acuity; but among colors their perception is clear only to yellow and blue. Red and green look alike and are dull, brownish, or achromatic. Yellowish greens and orange may appear like dull yellow. Bluish green and violet may resemble dull blue. A pure blue has considerably more brilliance than red or green.

Red-green blindness may be associated with shortening of the visible part of the spectrum at the red end. To such observers (protanopes), red appears dark, almost black. Light red is confused with slightly darker gray and still darker blue-green. In the case of red-green blindness where there is the normal extent of visible spectrum (called deuteranopes), green is confused with yellowish gray and magenta.

These types of color blindness are heritable and will follow laws of heredity. One compensation is that the afflicted person may have higher acuity in dim light, perhaps owing to a greater preponderance of rods in his eye. Women are spared, less than 1 per cent of them having the defect. In white males the percentage runs about 4. The percentage among Negroes is less than 3, among Indians, less than 2. These relatively few persons are obliged to make up for the deficiency by judging certain colors mainly in terms of value.

How Many Colors?

You may take your choice: either there are innumerable colors or there are very few—much depends upon attitude. If you were an average person, taking an average view of colors (as in a department store), if the illumination was average, if the samples before you were isolated from one another, and if you looked at them from a slight distance, the chances are that you would not be able to distinguish or recognize more than a few hundred different tones.

Frankly, the author holds to the contention that there are few colors essentially, and *very* few as far as average human interest and attention are concerned. There seems to be little reason for speaking of vast numbers, when it is so natural for most of us to look at colors in simple ways and to be little concerned over minor differences.

Plenty of estimates have been made. Edwin G. Boring has reviewed the work of several investigators. As to pure spectral hues, surprisingly few can be discerned by the average person. Von Kries placed the number at 208; Külpe at 150. In recent years, Selig Hecht wrote, "The normal eye can separate the visible spectrum with complete certainty into about 180 patches of hues which cannot be made to look like one another by varying their intensities." L. A. Jones put the figure at 128.

Pure spectral hues are, of course, only a small part of the world of color. Variations toward white, toward black, and toward gray account for many more. As to estimates of the total number of colors to be seen by the human eye, Titchener set a low figure of 32,820. Dorothy Nickerson and Sidney M. Newhall, after careful figuring, arrived at a total of 7,500,000. Deane B. Judd and Kenneth L. Kelly have mentioned "about ten million surface colors distinguishable in daylight by the trained human eye."

So I repeat, colors may be few or many, depending on the way a person goes about thinking of them and looking at them.

Color Organization

Organized systems of color and methods of color designation are useful equipment. Every practical worker in industry will recognize the problems of setting up standards, keeping notations, and otherwise preserving some method of control and order over the colors with which he deals. There are in existence today several convenient systems, charts, and instruments; a brief description of them should have value. One or more of these should probably be adopted and used by many of my readers.

A rather thorough list of American color standards will be found in Appendix B of this book. A fuller description of the more important ones is presented here.

Publications

The most venerable set of color standards in America is that of Robert Ridgway, *Color Standards and Nomenclature*. The last edition was published at Washington by the author in 1912. This book, which contains about 1,000 samples, each named by word, has been extensively used by naturalists and archaeologists.

In the field of horticulture, there is *The American Colorist*, written by the author of this book. This contains 12 process charts exhibiting some 500 different samples. Another source of reference is the *Fischer Color Chart*, containing 108 colors for the description and identification of flowers, and published by the New England Gladiolus Society. Very elaborate standards have been published in England. These are known as the *Horticultural Colour Charts*. They contain about 800 samples and are issued by the British Colour Council.

As to commercial standards, useful in business and industry, a very competent service is rendered by the Textile Color Card Association of New York. This organization also issues *The Standard Color Card of America*, Ninth Edition. The purpose of this is to standardize colors for the textile and allied industries. It contains some 216 samples of staple colors that have been widely accepted by consumers and manufactured by various industries.

In England, the British Colour Council publishes a set of charts containing 180 samples of dyed silk, extensively used by British industry to designate color names.

A Dictionary of Color

A very practical reference book—one that many of my readers should have—is *A Dictionary of Color*, by A. Maerz and M. Rea Paul (McGraw-Hill Book Company, Inc., New York). This contains over 7,000 samples. Essentially a book of names and colors, this is not a color system. A very scholarly job has been done in coordinating the terminology of many industries and arts and attempting to establish logical identifications. To anyone who occasionally wonders what a certain color should be called, this book will be of real help and will give him a meaningful term that has a sound basis in previous usage and identity.

The Munsell System

Today, there are two major systems of color that may lay claim to wide recognition, both in America and in Europe. These are the creations of Albert H. Munsell (1859–1918) and Wilhelm Ostwald (1853–1932).

They are greatly alike in several respects, and probably both should be owned by anyone active in the field of color.

The two systems have these elements in common: (*a*) an orderly sequence of colors that plot the entire realm of visual sensation; (*b*) a three-dimensional conception in which the general form of the solid comprises a white and a black apex, a neutral gray axis, pure hues about an equator, and intermediate scales of tints, shades, and tones in methodical arrangement; (*c*) a method of color identification in which each individual tone within the solid is to be measured and noted.

The color solid of the Munsell System follows the general shape of a sphere. However, because the pure colors do not have identical lightnesses or saturations, light colors like yellow are placed near the white apex, and dark colors like purple near the black. Again, colors of strong saturation, like pure red, extend farther from the neutral gray axis than colors like blue-green. These features are illustrated in accompanying diagrams and make the Munsell solid take somewhat the shape of a rhomboid with rounded corners.

For Munsell the three dimensions of color are hue, value, and chroma.

Hue is the quality that distinguishes red from orange, orange from yellow, etc. It is identified in the system with letters, as follows:

R: Red
YR: Yellow-red
Y: Yellow
GY: Green-yellow
G: Green
BG: Blue-green
B: Blue
PB: Purple-blue
P: Purple
RP: Red-purple

Each of these 10 major hues is given 10 steps—for a total of 100 colors throughout the complete circle.

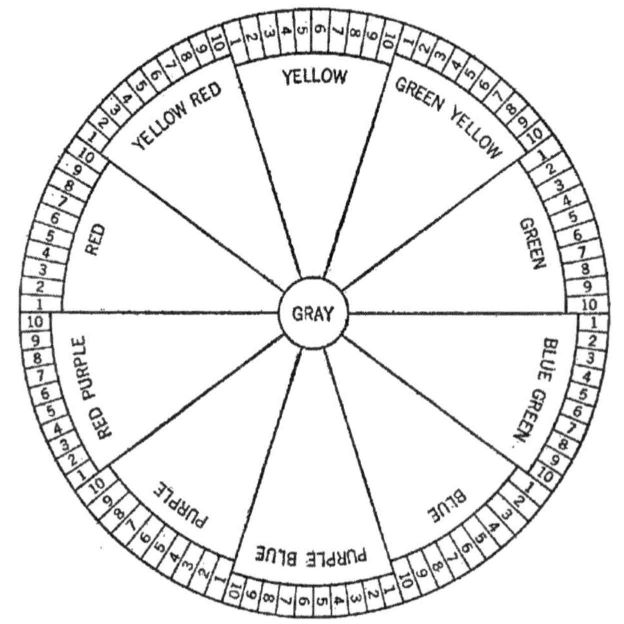

The Munsell color circle.

Value refers to apparent lightness as compared with white, grays, or black. Nine value steps from black to white have numbers from 1 to 9.

Chroma refers to purity, saturation, or apparent degree of departure from a "neutral" gray. A color of weak chroma will thus have a low number (1, 2, 3). Colors of strong chroma will have correspondingly high numbers, a brilliant red reaching a chroma of 14 or more.

Hence, hue is given a symbolic letter (sometimes preceded by a number to indicate its exact location on the Munsell color circle). The identification of value follows next, with a number; and the identification of chroma is third, with another number.

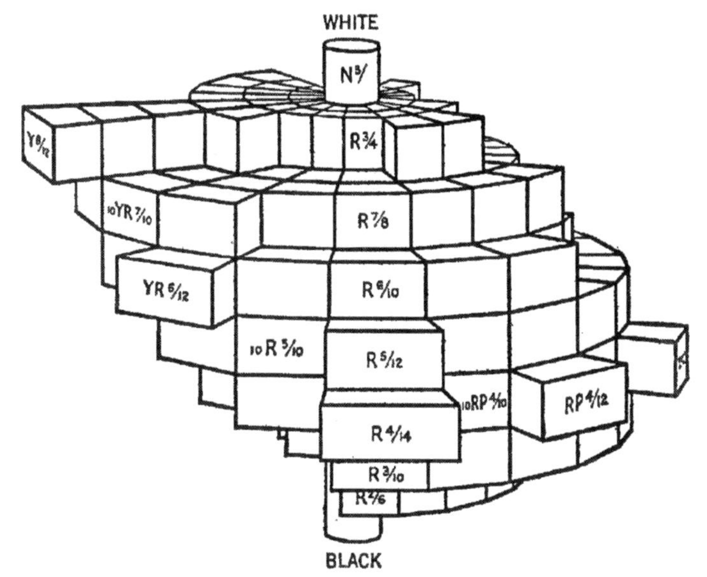

The Munsell color solid.

Here, then, are a few notations, duly translated:

Y 3/2. This would be a deep olive color, being a yellow equivalent in value to a deep gray (3) and weak in chroma (only two steps from neutrality).

10PB 7/6. This would be a lavender. The value (7) is equivalent to a fairly light gray; the chroma, six steps from neutrality, is clear and bright.

R 5/3. This would be a soft, grayish red, midway in brightness between black and white, and only three steps removed from gray.

The Munsell system holds a sympathetic place in the heart of many leading American scientists. Since 1912, work by the National Bureau of Standards, the Agricultural Marketing Service of the Department of Agriculture, and committees of the Optical Society of America has led to many inquiries and to the perfection of the system on many details. The *Munsell Book of Color*, as published today, contains 40

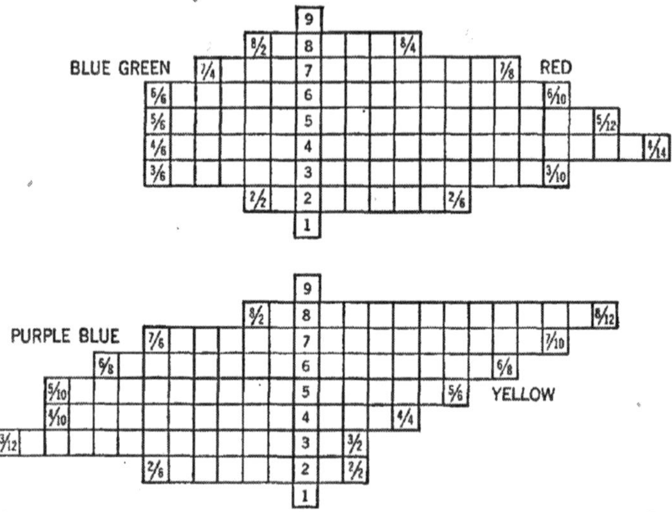

Sections of the Munsell color solid: red and blue-green; yellow and purple-blue.

charts. (Munsell Color Co., Baltimore.) It is unquestionably the most widely used method of color designation in America.

The Ostwald System

Whereas the Munsell system finds its dimensions in hue, value, and chroma, Ostwald is concerned with hue, white, and black. Following principles stated by psychologists such as Ewald Hering, Ostwald's system recognizes that every color seen by the eye is comprised of proportions of hue, white, and black, and that all such elements are measurable.

His solid, as illustrated, has the form of a double cone. His color circle, at the circumference, is made up of 24 hues, which are designated with numbers and letters, as follows:

1, 2, 3	Yellow	13, 14, 15	Ultramarine blue
4, 5, 6	Orange	16, 17, 18	Turquoise
7, 8, 9	Red	19, 20, 21	Sea green
10, 11, 12	Purple	22, 23, 24	Leaf green

Each of the 24 hues is made to comprise a monochromatic triangle having 28 tones. The neutral gray scale has eight

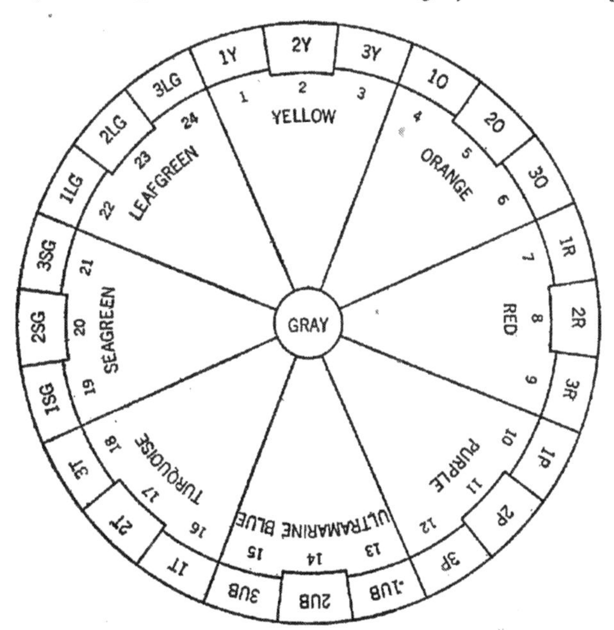

The Ostwald color circle.

steps and is identified with the letters *a* (for white), *c*, *e*, *g*, *i*, *l*, *n*, *p* (for black).

Every color, pure or modified, is to be described with a combination of these letters, and the letters indicate the pro-

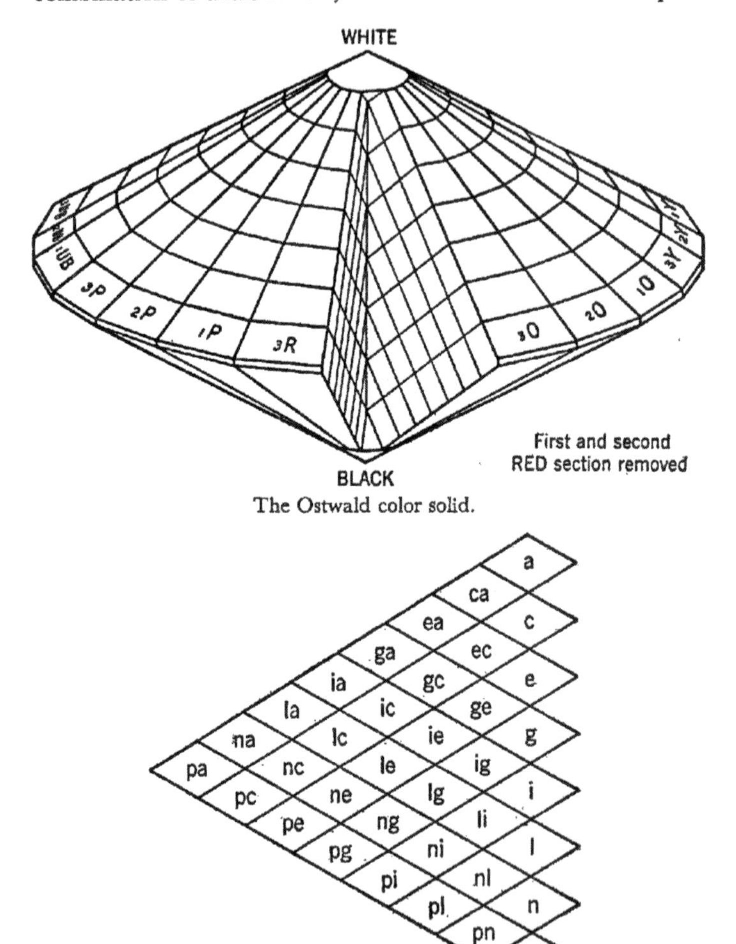

The Ostwald color solid.

Section of the Ostwald color solid.

portionate amounts of white and black contained in the tone. The following, for example, are a few designations. The numeral refers to the hue, as on the color circle. The first letter refers to the white content, and the second letter to the black content—both traced from corresponding amounts of white or black on the gray scale.

8 *pa:* This would be a relatively pure red. The first letter, *p* (black), indicating an absence of white in the mixture, and the second letter, *a* (white), indicating an absence of black.

2 *pi:* This would be a deep yellow or olive, the first letter, *p*, indicating an absence of white, and the second letter, *i*, a slight amount of black.

12 *ea:* This would be a lavender, with an *e* amount of white as on the gray scale, and without black.

8 *le:* This would be a soft grayish red, having an *l* amount of white and an *e* amount of black, as on the gray scale.

The Ostwald System is receiving increased recognition both in America and abroad. It has its limitations, however. Its end points, for example, are at all times fixed. (New and more brilliant chromas cannot be added, as they may on the Munsell "tree," without reforming the entire scheme.) Two sets of standards are available today. One is published by Winsor & Newton of London. The other, more elaborate, is issued by the Color Laboratories Division of the Container Corporation of America, Chicago.

The ISCC-NBS Method

There is one method of color designation which I feel should be far more widely known in America than it is. Some years ago, the Inter-Society Color Council, in collaboration with the National Bureau of Standards, studied the problem of devising a system of color description for drugs and chemicals to be used by the U. S. Pharmacopoeia. This system, known today as the ISCC-NBS Method, was "to be sufficiently standardized as to be acceptable and usable by science, sufficiently broad to be appreciated and used by science, art, and industry, and sufficiently commonplace to be understood, at least in a general way, by the whole public."

Briefly, the system is based upon a meaningful use of words. All but very grayish colors are described by hue names (red, green, blue, purple, etc.) preceded by modifiers, such as pale, light, brilliant; weak, moderate, strong, vivid; dusky, dark, deep (pale blue, moderate blue, strong blue, deep blue, etc.). For very grayish colors, hue names become modifiers of white, gray, or black (bluish white, medium bluish gray, bluish black, etc.).

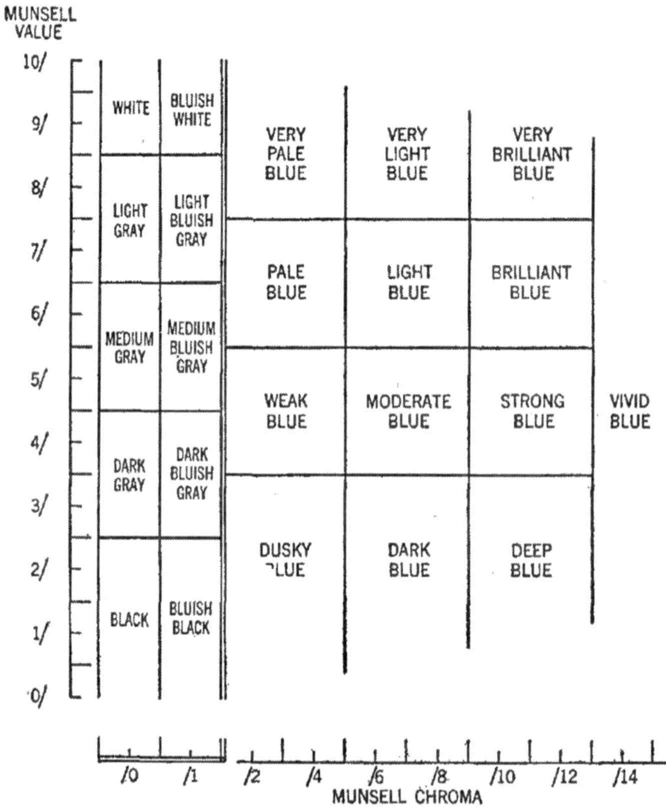

ISCC-NBS terms as applied to a Munsell color chart for blue.

This will all be rather clear in studying the accompanying diagram in which the ISCC-NBS method has been applied to a Munsell blue chart. In effect, it is possible to write fairly intelligible descriptions of color by this means. More standardization of the sort is highly desirable in many instances.

The Munsell System has been completely translated into ISCC-NBS terms, and the two together make an ideal pair. A descriptive booklet, available through the National Bureau of Standards at Washington, is known as Research Paper RP 1239, "Method of Designating Colors."

The ICI System

In the scientific specification of color through the use of instruments (colorimeters), the light emitted, transmitted, or reflected by an object or by a material is analyzed. Colorimetry gets down to fundamentals. A color is split up into its component spectral parts through the use of a prism, and the amount of each part is measured by photometry. The basic tool of modern colorimetry is thus the spectrophotometer.

Judd writes: "After taking light beams all to pieces, modern colorimetry puts them together again by a synthesis like that of the human eye. The result of putting the beam back together is to compute a specification of the color, just as the human eye synthesizes the spectral components of the beam by responding with a certain color sensation. These computational steps are based upon the tristimulus method of specifying color. It was found by Maxwell that most colors can be duplicated by shining three spot lights on a white screen—a red, a green, and a blue spotlight. The amounts of flux in these three lights constitute the specification of the unknown color. . . .

"Of course, the tristimulus specifications of each part of the spectrum have to be known. These specifications vary somewhat from observer to observer and may be given for any of a wide number of primary triads. It is almost universal practice, however, to use tristimulus specifications of the spectrum according to the standard observer and primaries recommended in 1931 by the International Commission on Illumination. The system of color specification so produced is known as the ICI system. The primaries used in the ICI system were chosen for the express purpose of simplifying the reduction of spectrophotometric data."

An accompanying chart shows a mixture diagram of the ICI system. The spectral quality of any color is represented by a point on this diagram. ICI designations are expressed in fractional form and are written with the letters x (for red), y (for green), and z (for blue). In analyzing a color, the proportion of each primary (x, y, or z) in the mixture is determined through instrumental means. (Since the fractional parts, x, y, z, always add to 1.0, only x and y really need to be plotted,

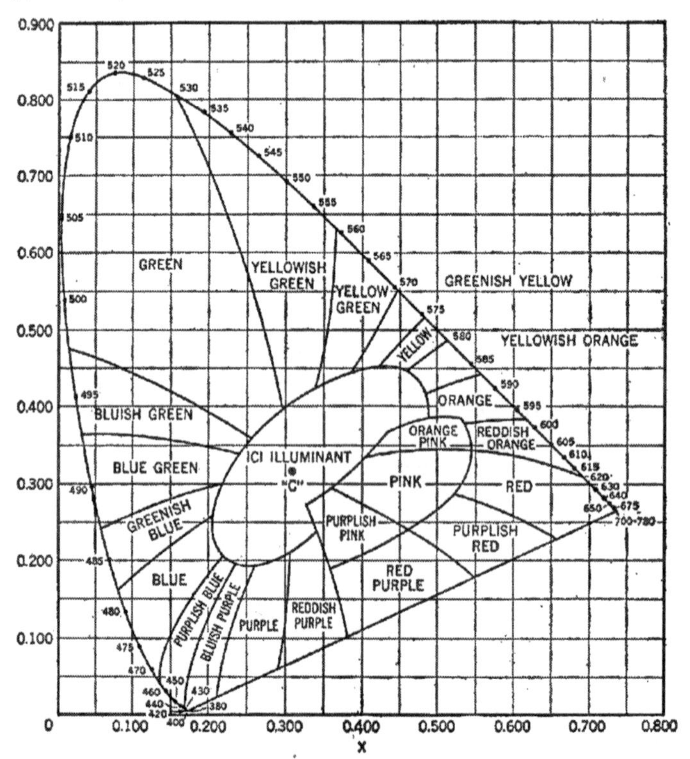

ICI mixing diagram. The color designations shown are those proposed by Kelly for lights. (*From Research Paper RP 1565, National Bureau of Standards.*)

z being the sum left over to equal 1.0.) Thus, a bright red printing ink may have the proportions x – 0.5610, y – 0.3175, z – 0.1215—the proportion of red (x) being predominant. (Note that the sum of these numbers is 1.0.) Though the principle of the ICI system is perhaps simple enough, the problem of color measurement is one of the most complicated to be encountered anywhere in the scientific world.

Colorimeters

A number of mechanical devices and optical instruments are available for more scientific and accurate measurements of color. An excellent review of them has been presented by K. S. Gibson in the *Journal of the Society of Motion Picture Engineers* for April, 1937. Some of these use material standards, chips, disks; others use spectrum primaries formed by prisms or diffraction gratings; and still others employ filters.

The Lovibond colorimeter makes use of colored glasses, for example, and is widely employed for the grading of liquids. In this method, red, yellow, and blue glasses of different depths are arranged to absorb white light and thus accomplish a match.

Bausch and Lomb has manufactured a disk colorimeter using Munsell standards. "The most serious drawback to the method would seem to be the uncertainty in the colorimetric values of the Munsell papers used as standards, either due to lack of initial certainty in the values or to change resulting from handling or other usage" (K. S. Gibson).

In colorimeters using spectrum primaries (Guild, Nutting, Priest), the color is matched either through a combination of primaries (red-green-blue) or more directly against a specific region of the spectrum (dominant wave length). Gibson writes, "In general, these instruments are more suited to research than routine testing or control purposes and have probably been used most in investigations of the colorimetric properties of the eye."

In filter photometers (Duboscq, Priest-Lange, Martens, Pfund), "the color of one side of a comparator field is adjusted, usually by varying the length of optical path through a solution, until it is brought to the nearest color match with the other side of the field, whose color is produced by a standard solution or filter."

Perhaps the best known and most distinguished instrument today is the Hardy Recording Spectrophotometer, manufactured by General Electric. This employs an electric eye and automatically draws the reflectance curve of a sample for all wave lengths throughout the visible spectrum. Although the instrument is quite expensive, readings from it may be had from several sources, such as the Electrical Testing Laboratories of New York. Wherever a permanent record of a color is desired, a spectrophotometric chart is desirable and very useful for future reference.

A Psychological Colorimeter?

The problems of color measurement have always troubled the author. While instruments, such as the recording spectrophotometer, serve an essential purpose in writing highly scientific notations, they seldom meet the more commonplace needs of industry.

The average form of colorimetry today may be described as an attempt by the physicist to adapt certain known facts about radiant energy to the more psychological responses of the human eye and mind. This is generally a complex, even a clumsy, process. The gap between physics and psychology has been well recognized. L. A. Jones writes that many in the field of physics have felt that "so long as color is defined as a sensation, it cannot be measured by physical instruments or by procedures reconcilable with physical principles of measurement. . . . The measurement of color, as it is practiced at the present time, is admittedly not a purely psychological measurement, nor, on the other hand, are such measurements purely physical."

Color specification in industry is commonly performed by skilled workers. They are, however, by no means versed in physics. They are men who must mix a color and prepare a formula. They may use paints, inks, dyes, textiles, ceramics, and a wide array of different materials. In colorimetry there is tremendous advantage in having the means to record the psychological appearance of any color, to specify it and describe it in definite terms, and to be able to see what it looks like at any time in the future.

Nearly all color matching is a process of skill and judgment in repeating what has been done before and in relying upon the worker's ability to effect a satisfactory duplication.

Most colorimeters in existence today thus fall short of the following counts:

1. They do not show the sample in its actual, psychological appearance.

2. The measurement taken is usually remote and foreign to the worker's own procedure in mixing colors.

3. The charts drawn require complex interpretation.

4. Few, if any, allowances are made for numerous factors that enter into the problems of color matching—luster, gloss, texture, top and under tones, etc.

5. The psychological elements of sensation—notably black—are not regarded.

6. The qualities measured, the manipulations of the dials, the ways in which notations are recorded, bear little comparison with the worker's own method and experience.

I mention these shortcomings—and end my book—having two thoughts in mind. First, I have been working on a "psychological" instrument myself. This is being designed to help clarify color specification in the simplest possible way, to give prominence to human factors, to avoid the complexities and irrelevancies of physics, and to offer industry an instrument that might readily be used and understood by any average person—specifically, the type of practical individual encountered in business fields. With luck on my side, a crude model, already built (and it works surprisingly well!) will be perfected, manufactured, and marketed one of these days.

Second, this whole business, art, and science of color has a long way to go. Great progress will follow a number of wise reconciliations. The aesthetics of styling will have more to do with scientific research techniques. And highly scientific things, such as colorimeters, will become more humble in conception and better adjusted to the singularities of human perception.

The hall of beauty and the hall of science will be less isolated from each other. They will be built closer together and will be connected by an arcade, crowded with people.

Appendix A

A REVIEW OF SALES RECORDS

THE following sales records have been gathered from a wide variety of sources through the efforts of American Color Trends and the Color Research Department of the Eagle Printing Ink Company. Wherever possible, the data are of a broad nature and are summaries of consumer preference as revealed in the reports of several, rather than few, manufacturers.

General consumer goods, mainly ot the more durable type, have been chosen for analysis. Women's fashions have been excluded because of the highly ephemeral nature of color trends.

The period included is generally that from 1940 to 1944, inclusive (this covering normal, as well as abnormal, markets brought about by war conditions).

Exterior Paints

Paint is a basic product in the study of general color trends. The figures herewith represent the summary of one large national paint dealer for all his grades.

Color	Per Cent
Total white	54.5
Total color	45.5

The popularity of various colors is given below:

Color	Per Cent	Color	Per Cent
Cream	16.7	Medium gray	8.1
Emerald green	15.8	Pearl gray	7.7
Ivory	12.1	Buff	7.7
Chocolate brown	11.2	Colonial yellow	5.3
Blind green	10.6	Glass black	4.8

Interior Paints

The same national paint dealer found the following totals for all grades of interior paints:

Color	Per Cent
Total white	30
Total color	70

The popularity of the colors was as follows:

Color	Per Cent	Color	Per Cent
Ivory	29.8	Ivory tan	6 7
Cream	13.2	Powder blue	4 6
Holland blue	9 7	Nile green	3.7
Peach	9.3	Sunshine yellow	3.0
Light buff	8.3	Pearl gray	2.7
Lettuce green	6.8	Dusty rose	2 4

Decorative Enamels

For decorative enamels in all grades, the following percentages held true:

Color	Per Cent
Total white	30
Total color	70

The colors ranked as follows:

Color	Per Cent	Color	Per Cent
Ivory	22.0	Medium green	5.9
Vermilion	11.5	Florentine blue	5.8
Cream	7.9	Canary yellow	4 0
Black	7.2	Golden oak	3.7
Old ivory	6.3	French gray	3.6
Lettuce green	6.3	Orchid	2 0
Simmons brown	6.2	Orange	1.7
Ivory tan	6.0		

Calcimine

In a line of hot-water calcimine, suitable for interior walls and ceilings, the consumer had the following color preferences.

Color	Per Cent	Color	Per Cent
Ivory	29.23	Light pink	2 52
Cream	18.80	Light green	2 02
Buff	10.94	Orchid	1 98
Pale buff	9.42	Pink	1 91
Tan	5.12	Sea green	1.26
Peach	4.63	Light blue	1 01
Yellow	3.60	Grayish green	0.50
Green	3.32	Neutral gray	0.43
Blue	2.92	Brownish gray	0.39

Asphalt Roofing

The preference for a white house with green trim is probably in large measure influenced by a similar preference for green roofs. From the records of one large manufacturer of asphalt shingles come these sales figures:

Color	Per Cent	Color	Per Cent
Bright green	28.19	Russet blend	3.25
Blue black	18.28	Gray-blue blend (texture)	2.73
Green blend	10.73	Warm brown blend	2.20
Blue blend	9.00	Slate green	2 06
Cool brown blend	7.36	Gray-green blend (texture)	2.03
Tile red	4 81	Gray-green	1.02
Red blend	4.76		
Blue-green blend	3 58		

Rugs and Carpets

We present here the results of consumer choices in 1940, 1941, and 1943, as shown by sales of one company in medium-priced plain broadloom.

1940 Color (Ranking Only)	1941 Color	1941 Per Cent	1943 Color	1943 Per Cent
1. Dusty rose	1. Dusty rose	10	1. Mauve	11
2. French blue	2. Light gray	10	2. Dusty rose	10
3. Jade green	3. Light blue	9	3. Dawn gray	8
4. Dawn gray	4. Jade green	7	4. French blue	7
5. Burgundy	5. Mauve	7	5. Jade green	7
6. Apple green	6. Apple green	7	6. Apple green	7
7. Castilian red	7. Deep red	7	7. Deep red	7
8. Jewel blue	8. Burgundy	5		
9. Ivory beige	9. Rose taupe	4		
10. Rose taupe	10. Blue-green	4		

The Institute of Carpet Manufacturers of America developed a very extensive report, which has bearing not only on color but on design as well and represents a cross section of the entire industry.

In general, floral and leaf patterns have been in greatest favor during the past few years and now comprise about half of all designs in Axminster, Wilton, and printed velvet rugs and carpets. Texture designs and Oriental patterns have decreased in popularity.

In all types of rugs and carpets, design types were distributed as follows during 1941.

Design	Per Cent	Design	Per Cent
Leaf, two-tone, and floral	47.3	Period and eighteenth century	4.8
Oriental	19.1	Modern	4.6
Texture	10.7	Colonial	4.4
Plain and Moresque	5.4	Chinese	3.7

As to ground colors, the institute report offers very comprehensive figures. We quote color percentages on all types of figured patterns and plain colors.

Color	Figured, Per Cent	Plain, Per Cent
Blue	19.7	19.8
Tan	15.6	12.3
Rose	14.8	10.6
Burgundy	11.1	7.1
Green	10.6	14.3
Red	8.3	6.8
Brown and wood	7.9	10.7
Rust	3.6	2.4
Taupe	2.5	5.3
Ivory and white	1.7	0.9
Peach	0.5	3.7
Black	0.5	0.8
Purple	0.2	2.1
Yellow and gold	0.2	3.2
Varicolor	2.8	

Regarding color trends, the institute report also includes data based on a survey among buyers and merchandise managers in retail stores. In the main, blue has been the color of greatest popularity. Rose has shown great rise in favor.

Declining colors have been burgundy and green. Tan has shown a slight increase, at the expense of rust.

Linoleum and Felt Base

Color preferences in linoleum are included here from the sales records of one large manufacturer.

Color	Heavy Weight, Per Cent	Standard Weight, Per Cent
Multicolored	10	27
White, plus	27	15
Blue	18	12
Tan	9	11
Green	7	10
Red	13	10
Gray	2	8
Black	14	7

In felt-base floor coverings, distinctly different color preferences are shown for floral types of patterns as against marbelized tiles. Floral goods are also predominantly in rugs, while marbelized tiles are sold chiefly by the yard. (Figures show percentages of sales in identical or closely similar patterns.)

Floral Rugs

Color	Per Cent
Tan	9.2
Blue	6.3
Beige	9.1
Blue	6.4
Blue	5.9
Rose	5.6
Red (Persian)	5.6
Tan	4.0

Tile Rugs

Color	Per Cent
Tan	7.4
Gray and black	7.1
Gray and red	6.2
Cream and green	5.3
Gray and blue	4.8

Marbelized Rugs

Color	Per Cent
Cream	4.2
Gray	3.7

Nautical Rug

Color	Per Cent
Blue	9.2

Floral Yard Goods

Color	Per Cent
Beige with pink	5.8
Blue with pink	4.6

Tile Yard Goods

Color	Per Cent
Beige with blue and red	4.6
Beige with red and green	4.3
Gray with red and blue	3.6
Tan	16.8
Gray and red	10.4
Cream and red	8.9
Gray and black	8.5
Gray, black and white	5.3

Marbelized Yard Goods

Color	Per Cent
Gray	7.3
Cream	6.9
Beige	6.7
Blue	6.3

Composition Tile

In one composition floor tile, having a *jaspé* effect, three tans sold 36 per cent of volume; two blues, 23 per cent; two reds, 16 per cent; white and black, 13.4 per cent; and two greens, 11.6 per cent.

A manufacturer of asphalt tile reports the following sales percentages:

Color	Per Cent	Color	Per Cent
Brown mottled	19.30	Solid reds	4.24
Tan mottled	17.57	Solid grays	2.30
Green mottled	13 46	Red mottled	2.21
Solid browns	12.30	White mottled	1.37
Solid black	9.36	Solid blues	0.95
Solid greens	6.41	Blue mottled	0.61
Gray mottled	5.26	Solid tans	0.33
Black mottled	4.29	Solid white	0.04

Though solid and mottled blacks were the least expensive of the tiles, they ranked fifth and eighth in consumer choice.

Wall Coverings

One manufacturer of wall linoleum reported "a very light salmon pink" as his most popular color, followed by "a white ground with veinings of red," with light buff third. Though he made no mention of other rankings, he did put green at the bottom of the list.

As to wallpaper, one manufacturer writes, "Eliminating ivory and tan backgrounds, the most popular colors in our wallpaper might be said to be soft rose colors, instead of dusty pinks, terra cotta peach was also popular. Blues still soft, as also the greens."

Upholstered Furniture

The following records on color preferences of upholstery are from a national manufacturer who makes goods in every price range.

Color	Per Cent	Color	Per Cent
Wine	17.7	Light green	3.0
Rose rust	11.0	Eggshell	3.0
Turquoise	7.2	Rose	2.7

Color	Per Cent	Color	Per Cent
Burgundy	6.5	Blue-green	2.5
Dark blue	6 1	Cedar rust	2.2
Royal blue	4.9	Rust	2.2
Maroon	4 8	Brown	1.5
Medium blue	4.8	Light blue	1.4
Red	4.7	Black	1.0
Dark green	4.0	Beige	0.8
Green-blue	3.6	Gold	0.4
Mauve	3.4	Gray	0.3

There are few important class distinctions in colors selected for furniture, except that the preference for wine is highest in low-priced upholstery, as it is for green, while in the upper brackets, rose rust and blue-green become more popular.

Outdoor Furnishings

Chairs used for porches and gardens sell best in the following colors—as reported by one manufacturer.

1. Red
2. Green
3. Blue
4. Orange
5. Yellow

Porch shades, which can be ornamental as well as useful, are sold in serviceable colors, as is shown by the report of one manufacturer.

Color	Per Cent	Color	Per Cent
Dark green	80	Light tan	5
Dark brown	10	Miscellaneous	5

Home Furnishings

From a department store and the headquarters of a large chain come general facts on color preferences in home furnishings, which will serve as a handy review to some of the statistics already quoted.

In general, the best sellers in home furnishings have been the rose family, the blue family, beige, natural, and green.

In items of home equipment and supplies, report gives these data:

Glassware

Color	Per Cent
Red predominating	45
Blue predominating	20
Orange predominating	15
Yellow predominating	10
Green predominating	10

Kitchen Items

Color	Per Cent
Red	70
Green	20
White and ivory	10

Bathroom Items

Color	Per Cent
White	85
Chrome	15

Kitchen ware: top seller, white with red trim.

Bathroom supplies: white with pearl-effect trim.

Kitchen furniture: white with red or black trim.

Upholstered furniture: rose tones first, beige second, burgundy third.

Mattress tickings: blue first, grays second, roses third.

Blankets

Here are the sales records of one large manufacturer of blankets for three years. The reduction in the number of colors offered has, of course, been a wartime measure.

Color	1941, Per Cent	1942, Per Cent	1943, Per Cent
Dusty rose	17.3	25.0	35
Blue	14 8	18.0	35
Winter rose	12 5	11.0	15
Green	10 4	9 5	15
Peach	12 8	5 5	
Beige	5.7	13 5	
Ivory or white	2.6	7.5	
Cedar or rust	7 5		
Dubonnet	6.6		
Deep blue	4.9		
Yellow	4.9		

Sheets

The bedroom preference for blue and pink is as evident in sheets as it is in blankets. The following figures were for 1940 and 1941:

Color	1940, Per Cent	1941, Per Cent
Peach	39	29
Blue	21	19
Green	19	19
Dusty rose	7	19
Yellow	9	12
Orchid	5	2

A manufacturer of inexpensive bordered pillow cases reported blue as his "best seller, with rose a close second All other colors, including green and orchid, are way down the list."

Mattresses

One maker of mattresses and box springs quotes interesting figures on two grades of mattresses. He remarks that where the consumer buys either a more expensive mattress or a set of twins (which naturally are double the cost of a single full size), a preference for gray seems to increase.

Grade	Blue, Per Cent	Rose, Per Cent	Gray, Per Cent	Green, Per Cent
$34.50				
Full size	32 3	28.9	18 6	20 2
Twin size	32.3	27.3	22.2	18.2
$39 50				
Full size	35.3	26.5	20.6	17.6
Twin size	34.6	24.5	22.6	18.3

He writes, "Blue is still the best color and this is holding fairly steady. Green is falling off, a bright rose and orchid have entirely disappeared."

Towels

Here is a brief statement by one manufacturer on color preferences for towels:

"In kitchen towels red is by far the best seller.

"In colored bordered terry towels we run goods on a basis of 35 per cent red, 25 per cent blue, 20 per cent green and the other 20 per cent including peach, yellow, and pink as most prominent."

Kitchen Ware

Popular kitchen colors are white, black, red, and green—year in and year out. Good primary colors that stay clean are apparently what the housewife has in mind when she buys kitchen towels, oilcloth, and other equipment.

In porcelain enamelware, white with red trim accounted for 34.5 per cent of sales, with black trim for 31.5 per cent, and ivory with green trim 34.2 per cent.

In lithographed bread boxes, canister sets, step-on cans, waste baskets, etc., white with red trim sold 52 per cent, with blue trim 27 per cent, and with green trim 21 per cent.

A higher quality line of decorative ware in solid pastel colors sold as follows: white 37.4 per cent, pastel yellow 18.6 per cent, pastel blue 17.5 per cent, pastel green 15.1 per cent, pastel rose 11.4 per cent.

In oilcloth, "white with red remains the preference, with green a fair second and blue a poor third."

Cotton Yard Goods and Trim

Fine-quality percale, sold for a wide variety of purposes, including domestic uniforms, children's garments, etc., had the following sales ranking.

1. Yellow
2. Peach
3. White
4. Black
5. Light blue
6. Powder blue
7. Eggshell
8. Tea rose
9. Midnight blue
10. Pink
11. Beige
12. Apple green
13. Turquoise
14. Royal blue
15. Gray
16. Tan
17. Dark green

It is interesting to compare these choices with those of bias tape, also available in a wide variety of colors, which might often be used to bind the yard goods.

Color	Per Cent	Color	Per Cent
White	30	Navy	2
Red	21	Nile green	2
Copen blue	11	Tomato red	2
Emerald	6	Seal brown	1½
Yellow	5	Orange	¼
Light blue	4	Old rose	¼
Pink	4	Dubonnet	¼
Black	3	Lavender	¼
Yale blue	3	Green	¼
Rose	3		

Plastics

The generous color ranges of modern plastics, and their almost countless uses, are indicative of consumer preferences in a wide variety of products, for plastics and color seem synonymous.

In sheet Cellophane, the top ranking color is red. Next in order are deep blue, deep green, light blue, pink, orchid, violet, black, and brown. Yellow and amber, however, are also big sellers, but mostly for their utility as wrappers, to retard rancidity in certain food products.

One manufacturer of molding powders, which are converted into finished products of all types, reports his best sellers to be white, then ivory, natural, red, tan, and blue.

From another source come figures on several different types of plastics.

Molding Powders

Color	Per Cent	Color	Per Cent
Red	24.0	White	10.1
Transparent	20.0	Blue	10 0
Green	13.3	Yellow and amber	9.3
Black	12.3	Brown	1 0

Plastic Sheeting with Pearl Effect

Color	Per Cent	Color	Per Cent
Silver	32.9	White	13.0
Green	16.5	Gold	7.8
Red	15.3	Blue	7.5
		Brown	7.0

Plastic Sheeting

Color	Grade A, Per Cent	Grade B, Per Cent
Transparent	28.4	77.8
White	21.4	7.8
Black	18.4	6.1
Red	15.7	3.0
Brown	10 0	0.7
Green	2 8	2 5
Blue	2 4	1.2
Yellow and amber	0.9	0.9

Plastic Tubing

Color	Per Cent	Color	Per Cent
Black	50.6	Blue	6 0
White	16.5	Green	4.4
Red	9 7	Yellow and amber	3 4
Transparent	7.6	Brown	1 8

Plastic Rods

Color	Per Cent	Color	Per Cent
Yellow and amber	27.0	Red	3.5
Transparent	24.9	Blue	1.3
White	23 9	Green	0 6
Black	18.8		

These figures may be somewhat confusing, because plastics are not sold to the consumer as such but as products, closures, etc., of all sorts. Yet, it is interesting to observe that the simpler hues find the widest sale, and that greatest volume is confined to few rather than many colors.

Automobiles

We include, finally, the sales records of a large maker of automobiles—covering the last year of production before the war.

Despite a trend toward light colors, visible in almost every commodity, black was the best seller. The demand for black, however, has been dropping steadily through the years. Green appeared to be rising in popularity; alone or in combination, it accounted for 27 per cent of output. Beige increased from 2.8 per cent to 3 per cent from 1940 to 1941. Gray, in shades and combinations, dropped from 36.1 per cent in 1940 to 23.7 per cent in 1941.

Colors	Per Cent	Colors	Per Cent
Black	26.0	Light and deep green	3.2
Light and deep gray	10.2	Light beige	3.0
Maroon	7.7	Pearl gray and deep blue	2.9
Two-tone olive green	7.7	Navy blue	2.4
Deep gray	5.9	Deep olive green	2 0
Pearl gray and black	4 7	Light olive green	1.6
Deep green	4 5	Maroon	1 6
Light and dark blue-green	4.4	Deep blue-green	1.4
Deep blue	3.7	Rust brown	0.9
Light blue-green	3 4	Other colors	4.8

Appendix B

CHECK LIST OF COLOR STANDARDS

THE list below represents a thorough compilation of color standards found in industry, science, and art. Some of these comprise general systems of color designation. Others are for specific products or for purposes of identification and safety.

AMERICAN COLORIST: Designed by Faber Birren and widely used in horticulture, art, and industry. Charts contain over 500 samples printed in 12 process charts. The Crimson Press, Westport, Conn.

AMERICAN STANDARDS ASSOCIATION: The most important recent standard for color identification is that issued by American Standards Association, New York, "Specification and Description of Color," Z44, 1942. Approved by a number of associations and manufacturers, it recommends the use of the recording spectrophotometer, a standard coordinate system; the Munsell System of material standards; and the Inter-Society Color Council method of designation.

ARTISTS' OIL PAINTS: Minimum standards to assure satisfactory color, working qualities, and permanence in oil paints used by artists. Adopted by a conference of manufacturers, distributors, and users. Commercial Standard, CS98-42, U. S. Department of Commerce, National Bureau of Standards, Washington, D. C.

BATHROOM ACCESSORIES: A group of seven colors for bathroom accessories: white, bath green, orchid, ivory, maize, bath blue, royal blue. Accepted by the National Retail Dry Goods Association and approved by a conference of manufacturers, distributors, and users. Commercial Standard,

CS63-38, U. S. Department of Commerce, National Bureau of Standards, Washington, D. C.

BAUMANN COLOR GUIDE: A card index system containing over 1,300 standards, with extra booklet showing complete samples in smaller size. The original edition was German, but an edition with English text has been produced. (Current source of distribution unknown.)

BRITISH COLOUR COUNCIL: A dictionary of 180 samples of dyed silk, used and accepted by British industry to standardize color names. Published at London.

CAMOUFLAGE COLORS: A series of nine colors, used as standards by the Army. Specification T-1213, Supplement A and B, U. S. Army, Corps of Engineers, Washington, D. C. Also, Color Card for Camouflage Finishes, issued by U. S. Air Corps. Contains eight samples, including insignia red, white, and blue. *Air Corps Bulletin* 21.

CAST STONE: A group of 14 colors for the finish of cast stone. Recommended Commercial Standard, CS53-35, U. S. Department of Commerce, National Bureau of Standards, Washington, D. C.

COLOR KIT: Designed by Faber Birren. Color identification is achieved through the use of disks and a mechanical spinning device. Numerical proportions are developed for all notations. The Crimson Press, Westport, Conn.

COLORFASTNESS: Standards for the testing of various textiles with reference to lightfastness and laundering are available for cotton goods, linen, pile floor coverings, rayon, silk, and wool. American Society for Testing Materials, Philadelphia.

DICTIONARY OF COLOR: Written by A. Maerz and M. Rea Paul. Shows over 7,000 samples, with color names based on historical origins and current usage. Published by McGraw-Hill Book Company, Inc., New York.

DU PONT COLOR CODE FOR SAFETY: Designed in collaboration with Faber Birren. System contains eight color standards, used for purposes of safety and identification in industry. E. I. duPont de Nemours & Co., Finishes Division, Wilmington, Del.

FISCHER COLOR CHART: A chart of 108 colors for the description and identification of flowers. New England Gladiolus Society, Norwood, Mass.

FLAG OF THE UNITED STATES: Scientific definitions of the red, white, blue, approved by all departments of the government. Section IV, Federal Standard Stock Catalog, Specification TT-C-591, July 3, 1934. Superintendent of Documents, Washington, D. C. (The Textile Color Card Association has prepared samples of these colors, also the flag colors of South American Republics.)

FOUNDRY PATTERNS ON WOOD: Standard system of marking wood foundry patterns, using black, yellow, and red, accepted by various associations. Commercial Standard, CS19-32, U. S. Department of Commerce, National Bureau of Standards, Washington, D. C.

FRENCH COLOR STANDARDS: Two widely used French sources for color standardization are the *Code Universeal des Couleurs*, by E. Seguy (720 samples), and the *Repertoire de Couleurs* of the Société Française des Chrysanthémistes.

GAS MASKS: A system to identify respirators and gas masks in industry. The colors used are white, black, green, blue, yellow, brown, red. *Bulletin* 512, U. S. Department of Labor, Bureau of Labor Statistics, Washington, D. C.

HILER COLOR CHART: A system of color harmony and color identification in chart form. Later edition exhibits 162 samples with card-index box containing mask matching apertures and showing mat and gloss finishes of the colors. Available through Favor, Ruhl & Co., Chicago and New York.

HORTICULTURAL COLOUR CHARTS: Published under the direction of the British Colour Council, London. Two volumes, containing about 800 color samples, used for designation in horticulture and industry.

INDUSTRIAL ACCIDENT PREVENTION SIGNS: Standards for the use of colors on signs in factories: red for danger signs, yellow for caution signs, green for safety instruction signs, black for directional signs, etc. Z35.1—1941, American Standards Association, New York.

INTER-SOCIETY COLOR COUNCIL: An effort to develop a standard designation for colors, using words commonly understood. Approved by various associations and applicable to widespread use in science, art, and industry. *Research Paper* RP1239, U. S. Department of Commerce, National Bureau of Standards, Washington, D. C.

KITCHEN ACCESSORIES: A group of six colors for kitchen accessories: white, kitchen green, ivory, delphinium blue, royal blue, red, accepted by the National Retail Dry Goods Association and approved by a conference of manufacturers, distributors, and users. Commercial Standard, CS62-38, U. S. Department of Commerce, National Bureau of Standards, Washington, D. C.

LUBRICATION OF MACHINERY: A proposed code (1944) for the use of color as an identification in the lubrication of machinery. There are five color standards for oils and three for greases, to be used on containers, fittings, etc. American Standards Association, New York.

MOTOR FUEL: Emergency method of test for U. S. Army motor fuel (all purpose). Standard consists of 4-ounce bottle of pinkish liquid used to make direct comparison with color of motor fuel. ES-32, American Society for Testing Materials, Philadelphia.

MUNSELL: The *Munsell Book of Color* is the most widely accepted system of color identification in the United States. The new edition, in two volumes, contains 40 charts and over 900 samples. Published by Munsell Color Co., Baltimore.

NATIONAL ELECTRICAL MANUFACTURERS ASSOCIATION: Standards have been established for the colors of flexible cords, glazed cotton braids, control cables, and type SN building wire (see Textile Color Card Association).

NU-HUE COLOR DIRECTORY: An elaborately boxed series of color standards in actual paint—over 1,000 samples, and all mixed from a basic palette of paint toners. Distributed by Martin-Senour Co., Chicago.

OSTWALD: The Ostwald System of color, widely used in Europe, England, and America, contains over 600 samples. Two series of standards are available, one produced by Winsor & Newton of London and New York, and the other by Container Corporation of America, Chicago.

PAINT MERCHANDISING COUNCIL: A comprehensive set of color standards derived from 10 basic paint toners, designed by Faber Birren, Clarence Deutsch, and the Paint Merchandising Council of Chicago. The system exhibits over 1,600 different colors in actual paint samples and comprises an atlas, a manual on color harmony, and a mechanical device for mixing and formulation.

PIPING IDENTIFICATION: Scheme for the uniform identification of piping systems in industry: red for fire protection; yellow (or orange) for dangerous materials; green (white, black, or gray) for safe materials; blue for protective materials; purple for valuable materials. Approved by American Standards Association, A13—1928.

PLOCHERE COLOR GUIDE: One of the most complete color systems ever produced in America. Contains over 1,000 samples, boxed and indexed. Available through G. Plochere, 1820 Hyperion Ave., Los Angeles.

POISONS, EXPLOSIVES, GASES: The Interstate Commerce Commission recognizes and insists upon special labels for shipments of gases, inflammable liquids, explosives, acids, etc. The colors generally used are white, black, red, yellow, blue, green.

PROCESS COLORS: Adopted (1927) by the Standardization Committee of the American Institute of Graphic Arts and approved by the American Association of Advertising Agencies and the National Association of Advertisers. Recommended standards for process red, yellow, blue, black.

RADIO MANUFACTURERS ASSOCIATION: Color standards have been established for fixed resistors (see Textile Color Card Association).

RIDGWAY: The most renowned historical work in America. Contains about 1,000 samples, each identified by name. Widely used by archaeologists and naturalists. Published by Robert Ridgway, Washington, D. C., 1912.

SANITARY WARE: Standard colors for plumbing fixtures and allied products made of vitreous china, enameled iron, etc. The colors are green, orchid, ivory, blue, light brown, black. Adopted by a conference of producers, distributors, and users. Commercial Standard, CS30-31, U. S. Department of Commerce, National Bureau of Standards, Washington, D. C.

SCHOOL BUS CHROME: Standard yellow adopted for school buses and approved by representatives of the 48 state departments of education. International Textbook Company, Scranton, Pa.

SCHOOL FURNITURE: Standard colors for school furniture, as adopted by a conference of producers, distributors, and buyers of school equipment. *Simplified Practice Recommendation*, R111-30, U. S. Department of Commerce, National Bureau of Standards, Washington, D. C.

SEDIMENTARY ROCKS: Goldman and Merwin Color Chart, showing 114 colors for the description of sedimentary rocks. Division of Geology and Geography, National Research Council, Washington, D. C.

SIGNAL GLASSES: Scientific description of colors used in signal glasses: red, yellow, green, blue, purple, lunar white. Signal Section Specification, 69-35, American Association of Railroads, Washington, D. C.

SOIL COLORS: A showing of 54 different colors of soil. *Miscellaneous Publication* 425, U. S. Department of Agriculture, Washington, D. C.

Textile Color Card Association: Seasonal color cards for the styling of consumer merchandise are issued at regular intervals to members. However, the Standard Color Card of America, 9th ed., contains 216 samples and represents staple colors having continual acceptance over the years in a wide variety of products and industries. This book contains probably the most important list of color standards in America today. It may be used for reference on the following standards: U. S. Uniform Colors; Official Colors of WAC; Colors of Ribbons for Decorations and Service Medals; U. S. Flag Colors; Army-Navy "E" Pennant Colors; Standards set up by National Electrical Manufacturers Association and Radio Manufacturers Association.

TRAFFIC DEVICES: Standard practices in the use of color on streets and highways—traffic lights, signs, curb markings, etc. American Association of State Highway Officials, Washington, D. C.

U. S. AIRCRAFT LINES: A chart showing color markings used on United States and British aircraft lines. This is an Army-Navy standard, T.O. No. 00-25-29 (restricted). It also includes a chart for hose markings designed to effect better standardization.

U. S. ARMY COLOR CARD: Army colors standardized for the different arms and services and approved by the Quartermaster General. Issued by Textile Color Card Association, New York.

U. S. Army, General Paint Specifications: A showing of 72 colors in various finishes, used by the Army in the purchase of paint and related materials. *Supplement* to No. 3-1, issued by the Quartermaster General, Washington, D. C.

U. S. Army Thread Color Card: Official color standards for olive drab, khaki, and drab sewing threads, with a list describing the use of each thread. Issued by the Quartermaster General, Washington, D. C.

U. S. Army-navy Aircraft Standards: A group of 15 colors used in the purchase of finishing materials for aircraft, Issued by the Bureau of Aeronautics, Navy Department. Washington, D. C.

U. S. Naval Medical Supply Depot: Standards have been established for the colors of slate green mosquito netting and wool blankets, dental smocks and dental towels, and Red Cross insignia for arm bands. U. S. Naval Medical Supply Depot, Washington, D. C.

U. S. Ribbons for Service Medals: General color specifications for the ribbons used on medals and decorations. U. S. Army, Specification 7-3B, Washington, D. C.

Annotated Bibliography

BECAUSE of the practical nature of this book and the fact that it has been written for a wide and varied audience, the text has not been interrupted by copious footnotes and references. The author doubts that the average reader would care to look up scores of books and articles, many of which would have but minor interest and value for him. Important sources, however, have been included in the text and in the following bibliography.

For anyone caring to assemble a useful library of books on color and related topics, the following list is recommended.

HISTORY: There are very few complete books on the history of color. To the author's knowledge, his own *Story of Color* is one of the most comprehensive and has much background material. Excellent reviews of color traditions in home decoration and women's fashions will be found in the two works of Elizabeth Burris-Meyer.

Faber Birren, *The Story of Color*, Crimson Press, Westport, Conn., 1941.

Elizabeth Burris-Meyer, *Historical Color Guide*, William Helburn, Inc., New York, 1938.

———, *This Is Fashion*, Harper & Brothers, New York, 1943.

M. Luckiesh, *The Language of Color*, Dodd, Mead & Company, Inc., New York, 1920.

Martin Lang (Faber Birren), *Character Analysis through Color*, Crimson Press, Westport, Conn., 1940.

V. Wheeler-Holahan, *Boutell's Manual of Heraldry*, Frederick Warne & Co., Inc., New York, 1931.

HARMONY: This is the most popular of all subjects for writers on color, and one that lends itself to a number of pet theories and notions. While dozens of references could be given, the following list includes the best among them. The finest review of color harmony is that of Klein. The author's *Monument to Color* expresses many advanced principles derived from the science of psychology. Hiler's book is beautiful and impressive. The other works are good standard references, widely known over the years.

Faber Birren, *Color Dimensions*, Crimson Press, Chicago, 1934.
———, *Monument to Color*, McFarlane, Warde, McFarlane, New York, 1938.
Fiatelle, *The Color Helm* (a chart), Fiatelle, Inc., Ridgewood, N. J.
Hilaire Hiler, *Color Harmony and Pigments*, Favor, Ruhl & Co., Chicago and New York, 1941.
Adrian Bernard Klein, *Colour-Music, the Art of Light*, The Norman W. Henley Publishing Company, New York, 1930.
Albert H. Munsell, *A Color Notation*, Munsell Color Co., Baltimore.
Wilhelm Ostwald, *Colour Science*, Parts I and II, Winsor & Newton, London and New York, 1933.
Arthur Pope, *The Painter's Modes of Expression*, Harvard University Press, Cambridge, Mass., 1931.
Walter Sargent, *The Enjoyment and Use of Color*, Charles Scribner's Sons, New York, 1923.

MERCHANDISING: There are not many books to be found in this particular field of color. The most comprehensive is that of Elizabeth Burris-Meyer. The problems of color in merchandising, however, are frequently presented in various trade magazines.

Elizabeth Burris-Meyer, *Color and Design in the Decorative Arts*, Prentice-Hall, Inc., New York, 1935.
M. Luckiesh, *Light and Color in Advertising and Merchandising*, D. Van Nostrand Company, Inc., New York, 1923.
———, *Color and Colors*, D. Van Nostrand Company, Inc., New York, 1938.
Louis Weinberg, *Color in Everyday Life*, Moffat, Yard & Co., New York, 1918.

PSYCHOLOGY: Almost any good textbook on psychology will be found to devote considerable space to color. The following references, however, are of particular interest. Godlove and Laughlin present a brief and popular review. Boring's work is general. That of Katz is particularly concerned with color constancy. Jaensch deals with the strange phenomenon of eidetic imagery.

Edwin G. Boring, *Sensation and Perception in the History of Experimental Psychology*, D. Appleton-Century Company, Inc., New York, 1942.

I. H. Godlove and E. R. Laughlin, *The Psychology of Color*, Technical Association of the Pulp and Paper Industry, 23: 473-525, June, 1940.

E. R. Jaensch, *Eidetic Imagery*, George Routledge & Sons, Ltd., and Kegan Paul, Trench, Trubner & Co., Ltd., London, 1930.

David Katz, *The World of Colour*, George Routledge & Sons, Ltd., and Kegan Paul, Trench, Trubner & Co., Ltd., London, 1935.

INDUSTRY: The problems of color and illumination in industry find their best exposition in the work of Luckiesh. Because books are scarce on this subject, several pertinent magazine articles are included in the following list.

Faber Birren, *Functional Color*, Crimson Press, New York, 1937.

———, "Color for Production," *Architectural Forum*, July, 1942.

———, "Color-conditioning in Modern Industry," *Dun's Review*, July, 1942.

———, "Color in the Plant," *Factory Management and Maintenance*, February, 1945.

M. Luckiesh, *The Science of Seeing*, D. Van Nostrand Company, Inc., New York, 1937.

———, "Brightness Contrasts in Seeing" (Written with Frank K. Moss), *Illuminating Engineering*, June, 1939; "Brightness Engineering," *ibid.*, February, 1944.

Parry Moon, "Wall Materials and Lighting," *Journal of the Optical Society of America*, December, 1941; "Reflection

Factors of Floor Materials, *ibid.*, April, 1942; "Colors of Furniture," *ibid.*, May, 1942.

VISION—THE EYE: Great books on vision and the anatomy of the eye have recently been published. The works of Polyak and Walls are remarkable achievements. The Bausch & Lomb item is illustrated with a series of striking full-color drawings printed on transparent sheets and is unusual in format.

The Human Eye in Anatomical Transparencies, text by Peter C. Kronfeld, Historical Appendix by Stephen L. Polyak, illustrations by Gladys McHugh, Bausch & Lomb Press, Rochester, N. Y., 1943.

Stephen L. Polyak, *The Retina*, University of Chicago Press, Chicago, 1941.

James P. C. Southall, *Introduction to Physiological Optics*, Oxford University Press, New York, 1937.

Gordon Lynn Walls, *The Vertebrate Eye*, Cranbrook Institute of Science, Bloomfield Hills, Mich., 1942.

SCIENCE: Textbooks and popular works on physics contain many references to the scientific aspects of color. In the following list, Luckiesch's book is something of a minor classic and belongs in every color library.

Sir William Bragg, *The Universe of Light*, The Macmillan Company, New York, 1934.

M. Luckiesh, *Color and Its Applications*, D. Van Nostrand Company, Inc., New York, 1921.

International Printing Ink Corp., *Three Monographs on Color:* "Color Chemistry," "Color as Light," "Color in Use," New York, 1935.

H. D. Murray and D. A. Spencer, *Colour in Theory and Practice*, American Photographic Publishing Company, Boston, 1939.

COLORIMETRY: For a complete list of commercial color standards and color systems the reader should consult Appendix B of this book. Perhaps the best known work on colorimetry today is that of Hardy. There are not many books on this subject. However, a very significant contribution is now in process under the direction of the Colorimetry Committee of the Optical Society of America. Introductory chapters of an exhaustive treatise have already been released, as is noted below.

A. C. Hardy, *Handbook of Colorimetry*, Technology Press, Massachusetts Institute of Technology, Cambridge, Mass., 1936.

K. S. Gibson, "The Analysis and Specification of Color," *Journal of the Society of Motion Picture Engineers*, April, 1937.

Report of Colorimetry Committee: see *Journal of Optical Society of America*, October, 1943; April and May, 1944; etc.

American Society for Testing Materials, *Symposium on Color—Its Specification and Use in Evaluating the Appearance of Materials*, March, 1941.

Technical Association of the Pulp and Paper Industry, *Symposium on Spectrophotometry in the Pulp and Paper Industries*, February, 1940.

American Ceramic Society, *Symposium on Color Standards and Measurements*, November, 1941.

INDEX

A

Accident-prevention signs, 224
Adler, W. F., 155
Advertising, 104–115
  case histories, 111–114
  color technique, 114–115
  direct-mail, 104–108
  revenue for color in magazines, 109
  space, 108–115
  and styling, different approaches required, 42–43, 88–89
Affective values of color, 27, 34
  on human senses, 35–36
Afterimage, 98, 99
  advantages of, in machine painting, 133
Albino plants, 153
Allah, 10
Allen, Frank, 36
Aluminum foil, used to control temperature, 142–143
American Color Trends, 50, 66, 67, 209
American markets, 45–46
American Society for Psychical Research, 157
American Society for Testing Materials, 223
American Standards Association, 222, 224, 225, 226
Architecture, Greek, 11
Arens, Egmont, 66, 148
Aristotle, 16
Artists' oil paints, standards, 222
Association of National Advertisers, 108
Attention-value, 96–97, 114
Aura, human, 164
Avicenna, 160

B

Babbitt, Edwin D., 161
Background, effect on color preference, 23, 25
Bagnall, Oscar, 164
Barnacles, dislike of brightness, 143
Bathroom accessories, standards, 222
Baumann Color Guide, 223
Beethoven, 170
Benham's disk, 90, 91
Benson, Dr. John, 162
Berthoff, 30
Birds, effect of color on, 32–33
Bissonnette, T. H., 164
Black, a definite color, 90
Blind flying, use of color in, 155–156
Blondes (blonds) color preference of, 37
  suitable colors for make-up and dress, 184
Blue, bactericidal properties, 162
  color preference of adults for, 39–40
  color therapy of, 162
  effect on the insane, 167
  glorification of, 82
  meaning in character analysis, 188
  in safety practice, 134
  suggests form of circle, 175
Blue-green, meaning in character analysis, 187
Bohr, 191
Boric acid, colored for safety, 154–155
Boring, Edwin G., 194

Boston Symphony Orchestra, 170
Bouguer, 100
Boyd, Dr., 146
Bragg, Sir William, 191
Brahma, 10
Brighouse, Gilbert, 166
Brightness contrast, 99
British Colour Council, 195, 223, 224
Brown, meaning of in character analysis, 188
Brunettes (brunets) color preference of, 37
suitable colors for make-up and dress, 184
Buddha, 10
Burris-Meyer, Elizabeth, 182, 183

C

California, southern, color preferences, 60
Camera, different from human eye, 100
Camouflaged colors, standards, 223
Carnegie Institution, 153
Cast stone, color standards, 223
Casual seeing, 133–135
Celsus, 160
Character analysis through color, 185–189
Chickens, effect of color on, 150
Children, color preferences of, 21, 24
Chopin, 170
Claypool, Dr. L. L., 153
College Inn Food Products Company, 153
Colleges and universities, color symbolism of major faculties, 180
Colonial period, use of color in, 183
Color, affective values of, 27, 34
attraction to insects, 31, 149–150
vs. black and white in advertising, 105
and character analysis, 185–189
and culture, 8
and design, 5
vs. design, 6
definition of, 17
Color, effect on flies and mosquitoes, 149–150
effect on judgment of time, weight, and distance, 165
elements of, 17
emotional appreciation of, 5
as energy and as sensation, 89
and form, 174–175
functional, 127
and growth of plants, 31–32
history of, 8–11
for identity, 114
and language, 13–15, 175–176
in medicine, 10, 146, 160–166
and mental telepathy, 157–158
and moods, 176–177
and music, 170–171, 174
and period styles, 182–184
and personality, 184–185
physical nature of, 190–192
physical reactions to, 163–166
of planets, 9
psychological types, 89
and religion, 10
specification of, 194–208
in treatment of insanity, 166–167
and war, 148–149
warm and cold qualities, 169–170
Color blindness, 192–194
frequency of, 193
Color combinations, preference for, 24–25
Color conditioning, 128–139
Color constancy, 94, 100–102
under chromatic light, 101
noted in hens, 100–101
Color contrast, 98–99
Color coordination, 51–54
manufacturers, 53
stores, 52
Color diffusion, 99
Color effect vs. color scheme, 85–87
Color engineering, 126–139
advantages of, 130
Color forms, 19–20, 78
Color harmony, 79–87
analogous, 79, 81
balanced, 80, 81

Color harmony, complementary, 80, 81
intermediate colors, 81
modified colors, 81, 84
pure colors, 81
Color organization, 194–195
Color preferences, 20–28, 39–40
of adults, 22-27, 39–40
of American Indians, 21
of blonds, 37
blue preferred by adults, 39–40
of brunets, 37
of children, 21, 24
for color combinations, 24–25
in commercial products of the past, 44
effect of background, 23, 25
effect of sunlight, 37–39
of Filipinos, 21
of high-fashion markets, 2–3, 85
of infants, 21
of the insane, 21
light colors vs. dark colors, 26
mass markets, 4–5, 11–13, 85
of men, 22
for modified color forms, 25–28
by nationality, 40
of Negroes, 21
pure colors vs. gray colors, 26
by regions in United States, 37–39, 60
reasons for, 29
sectional, for commercial products, 60
for single colors, commercial products, 76–77
of women, 22
Color research, 48, 63–73
basic principles, 63–65
benefits, 70–71
blotting papers, 96
deciding upon new colors, 65–66
packaging labels, 117
Color scheme vs. color effect, 85–87
Color specification, 194–208
Color standards, 222–228
Color styling, 11-13
few colors desired, 56–57
Color styling, high-fashion markets, 2–3
mass markets, 4–5
Color-styling departments, manufacturers, 53
stores, 50
Color symbolism, 177–182
Freemasonry, 179
heraldry, 179–180
Roman Catholic Church, 177–178
stock shows, 180
universities and colleges, 180
Color systems, 75–76
Munsell, 20*n*, 27, 75, 196–199
Ostwald, 75, 199–202
Color terms, 20*n*.
Color trends, 12, 57–59
causes of, 46–47
Color triangle, 18, 77, 84
Color vision, 192–194
affected by disease, 163
of birds, 33
of fish, 33
of insects, 30–31
of lower organisms, 30
of monkeys, 33
in twilight, 92
of vertebrate animals, 30
Colored illumination, 101
in displays, 120
in stores, 122–123
Colorfastness, standards for, 223
Colorimeters, 206–208
psychological, 207–208
Colors, and edible qualities in foods, 125
elementary, 18
and illusions of size, 97–98
on machinery, 133
of months and seasons, 181
to prevent rancidity, 144
primary, 18
recognizable by name, 14
Columbia University, 157
Cones and rods of eye, 91
Confucius, 10
Consumer research, 50, 66–70
benefits of, 70–71

Consumer reseach, consumer panels, 70
mail questionnaires, 70
manufacturers, 53
negative, 66–67
positive, 67–69
stores, 51
Container Corporation of America, 202
Continental Lithograph Corporation, 145
Coolness and warmth in color, 169–170
Cosmetics, origin of, 8
Cotton, grown in color, 152
Creslas, A., 32
Critical seeing, 131–133
Culture and color, 8

D

Dalton, John, 193
Daylight illumination, artificial, 94
Design, and color, 5
vs. color, 6
Deutsch, Clarence, 225
Deutsch, Felix, 167, 168
Dictionary of Color, 195, 223
Dimension in color, 97–98
Direct-mail advertising, 104–108
circulars, 105
envelopes, 105
letterheads, 105
reply cards, 106
rules for color, 107–108
Direct Mail Research Institute, 104
Displays, 119–120, 123–125
merchandise, 123–125
Dress shop, color scheme for, 123–124
Du Pont Color Code, 133–135, 223
Du Pont de Nemours and Co., E. I., Color Code for Safety, 133–135, 223
Dulux white for tanks, 141

E

Eagle Printing Ink Company, 104, 108, 116, 209
Eggs, control of color of, 151
Egner, Frank, 107
Egyptians, 8
use of color in medicine, 160
Eidetic imagery, 171
Electrical Testing Laboratories, 207
Electromagnetic spectrum, 191
Elementary colors, 18
Elements, ancient conception of, 9, 16
Evaporation, control of through color, 141
Extrasensory perception, influenced by color, 58
Eye, human, 91–92
dark-adapted, 93
focus of, affected by color, 97
to purple, 40
similar to camera, 91
Eyestrain, 129–130
causes of, 129
results of, 129–130
Eysenck, H. J., 21, 22, 24

F

Factory decoration, 128–139
case histories, 136–138
psychological consideration in, 135–136
Fashion, 71–73
Fergusson, James, 9
Ferree and Rand, 93
Filipinos, color preference of, 21
Finsen, Neils, 161
Fish, color change in, 100
effect of color on, 33
Flies, effect of color on, 149–150
Flint, Lewis G., 32
Floor coverings, hard surface, experience in color styling of, 68
sectional preferences of, 60
Florida, color preferences, 60
Fluorescein, medical use of, 146–147
Fluorescence, 145–147
Fluorescent light, 94–95
Fluorescent materials, 146–147
Foods, edible colors, 125, 173
prevention of rancidity in, 144
Ford, Henry, 1

Fordney, Major, 140
Form, and color, 174–175
mental appreciation of, 5
Foundry patterns, color standards, 224
Four quarters of earth, symbolism of, 8, 9
Foveal vision, 91
France, Anatole, 34
Franzen, Raymond, 117
Freeborn, S. B, 149
French color standards, 224
Fruit, control of color, 154
Functional color, 127
advantages of, in industry, 130

G

Gale, Ann Van Nice, 24
Galen, 160
Gallup, 66
Galton, Francis, 170
Garth, T. R, 21
Gaw, George D., 104
Gay Nineties, 73
General Electric Company, 154, 206
Gibson, K. S, 206
Godlove, I. H, 83
Godlove's principle, 83
Goldstein, Kurt, 34, 35, 165
Gore, John, 183
Greek architecture, 11
Green, glorification of, 82
meaning in character analysis, 187
in safety practice, 134
suggests form of hexagon, 175
Grisewood, E. N, 30
Guilford, J P., 21, 24–29, 79, 82, 89
Gundlach, E. T., 110

H

Hale, Philip, 170
Hardy, E., 149
Hardy, Le Grand H., 190, 192
Hardy Recording Spectrophotometer, 206
Hatchings of heraldry, 179
Heat absorption, use of paints, 155
Heat radiation, 140–143
in radiators, 142
in storage tanks, 141
Hecht, Selig, 18, 194
Helmholtz, Hermann von, 90
Helson, Harry, 101
Heraldry, 179–180
hatchings, 179
Hering, Ewald, 121, 199
Herodotus, 9
High-fashion markets, 2–3, 59
preference for color, 85
Hiler Color Chart, 224
Hindu Upanishads, 9
History of color, 8–11
Hoffman, 165
Hollywood, influence on color of women's apparel, 71
Homes, average number of colors in furnishings, 49
Hoodless, 150
Hooke,190
Hoover, W. H., 32
Hopi, color symbolism, 9
Horticultural color charts, 195, 224
Hospital decoration, 126
Howat, R. Douglas, 162
Human aura, 160
Human eye, 91–92
dark-adapted, 93
focus affected by color, 97
focus to purple, 40
similar to camera, 91

I

Ice, colored for skating rink, 156
ICI system of color designation, 204–206
Illumination, colored, 101
in displays, 120
in stores, 122–123
Imada, M., 24
India, original four castes, 10
Indians, American, color preference of, 21
symbolism, 8–9
Industrial design, 126–127, 138–139
Infants, color preferences of, 21

Insane, color preference of, 21
Insects, attraction to color, 31
color vision of, 30–31
Interior decoration, 120–126
commercial, 120–125
in factories, 132–138
for hospitals, 126
ideal dress shop, 123–124
restaurants, 125
in stores, 122–123
Inter-Society Color Council, 202, 222, 224
Interstate Commerce Commission, 226
ISCC-NBS method of color designation, 202–204

J

Jaensch, E. R., 34, 37, 171
Jeans, Sir James, 192
Jewelry, origin of, 8
Jones, L. A , 194, 207
Judd, Deane B., 149, 194, 204

K

Kansas State College, 151
Karwoski, T. F., 174
Katz, David, 100
Katz, S. E., 21
Kelly, Kenneth L., 194, 205
Kirsch, Fred, 150
Kitchen accessories, color standards, 225
Klüver, H , 33, 171
Kravkov, S. V., 35
Külpe, 194

L

Lang, Martin, 185
Lange, Dr , 146
Language and color, 13–15, 175–176
Lashley, K. S., 33
Legibility, 95–96
Light and life 30–31
Liszt, 170
Lodge, O. C., 149
Logan, H. L., 131
Lord, symbolized by blue, 10
Lubrication of machinery, color standards, 225
Luckiesh, M., 40, 92, 129, 131, 132, 138, 147, 163
Ludwig, 162
Luminescent paints, 155

M

MacDonald College, 151
Machinery, use of color on, 133
Maerz, A., 195, 223
Magenta, effect on insane, 167
Management (business), responsibility in matter of color, 61
Marie Antoinette, 182
Market data, distribution of total U.S. consumer budget, 45–46
number of U.S. dwellings, 45
occupations and trades, 45
retail sales, 46
total U.S. population, 45
urban and rural distribution, 45
Maroon, meaning in character analysis, 185
Mass markets, 4–5, 45–46
preference for color, 85
styling for, 11–13
Mauve decade, 183
Maw, Dr. W. A., 151
Maxwell, James Clerk, 17, 204
Meat, control of color of, 151
Medicine, uses of color in, 10, 146, 160–166
in diagnosis, 162–163
in healing wounds, 165
modern attitude, 160–161
in treatment of insanity, 166–167
in treatment of tumors, 165
Meloy, Thomas, 117
Menju, Dr., 164
Mennonites, 183
Mental telepathy and color, 157–158
Merchandise, displays, 123–125
unsold due to color, 50, 52, 61, 71

Mercury light, 92
Mercury vapor, detected by violet light, 154
Metzer, 163
Metzger, 35
Michelson, 192
Middle West (U.S.), color preferences, 60
Milk secretion, effect of color on, 164
Mizutani, Dr., 165
Modified colors, preference for, 25–28
Moleschott, 162
Montgomery Ward, 109
Months, colors of, 181
Moods and color, 176–177
Moon, Parry, 79, 80, 82, 131
Morgan, Dr. Agnes Fay, 154
Morley, 192
Morrison, Beulah M., 25
Mosquitoes, effect of color on, 149–150
Mozart, 170
Munsell, Albert H., 196
Munsell system, 20*n*, 27, 75, 196–199, 203, 204, 222, 225
Murphy, Gardner, 157
Murray, H. D., 25, 169
Music and color, 170–171, 174, 180
Musicians, reactions to color, 170–171

N

Napoleon, 182
National Electrical Manufacturers Association, color standards, 225
National Retail Dry Goods Association, 222, 225
Navahos, color symbolism, 8–9
Nebuchadnezzar, Temple of, 9
Negroes, color preference of, 21
Nelson, Lord, 149
New England, color preferences, 60
New England Gladiolus Society, 195, 223
Newhall, Sidney M., 75, 169, 194
Newstead, R., 149
Newton, Sir Isaac, 16, 190
New York, metropolitan, color preferences, 60
New York, metropolitan, influence on color of women's apparel, 71
Nickerson, Dorothy, 194
Northwest (U.S.), color preferences, 60
Nuttall, G. H. F., 149

O

Occidental College, 166
Odbert, H. S., 174
Orange, meaning in character analysis, 186
  in safety practice, 134
  suggests form of rectangle, 174
Ostwald, Wilhelm, 196
Ostwald system, 75, 199–202, 225

P

Pacific Coast, color preferences, 60
Package design, 102–103
Packaging, 116–119
  case records, 117
  writing ideal specifications, 118–119
Paint Merchandising Council, 185, 225
Paints, used for heat absorption, 155
Pancoast, S., 161
Paris, influence on color of women's apparel, 71
Paroptic perception, 34
Parsons, J. H., 17
Paul, M. Rea, 196, 223
Period styles, 182–184
Peripheral vision, 91
Perry, L. J., 149
Personality and color, 184–185
Pfeiffer, N., 32
Physics and color, 190–192
Piccard, 140, 141
Picton, 161
Pink, meaning in character analysis, 186
Piping identification, color code, 226
Planck, 191
Planets, color of, 9
Plants, effect of color on growth of, 31–32, 161
Plochere Color Guide, 226

Ponza, Dr., 166
Popp, H. W., 32
Population, urban and rural distribution, 45
Porter, L. C., 31
Prideaux, G. F., 31
Primary colors, 18
Process colors, standards, 226
Psychiatry, 166–169
Psychology of color, 159–166
  in advertising, 114–115
  in factory decoration, 135–136
Public taste in color, 6, 12
Purdue University, 152, 153
Purple, difficult to focus, 40
  meaning in character analysis, 188
  suggests form of oval, 175
Pythagoras, 160

Q

Quarters of earth, symbolism of, 8, 9

R

Radiation sense of human body, 35, 165
Radio Manufactures Association, color standards, 226
Raff, 170
Realism in color, 114
Red, color therapy of, 161–162
  glorification of, 82
  meaning in character analysis, 185
  in safety practice, 134
  suggests form of square, 174
Reeves, Cora, 33
Regional color preferences, 37–39, 60
  for hard-surface floor coverings, 60
Religion and color, 10
Research, 48, 63–65
  advertising, 104–114
  benefits of, 70–71
  direct-mail, 104–108
  packaging, 117–118
  space advertising, 108–115
  Starch reports, 108, 109
Restaurants, color scheme for, 125
Retail sales, total U.S. expenditure for, 46
Ridgway, Robert, 195, 226
Rimsky-Korsakoff, 171
Rockefeller, John D., 183
Rods and cones of eye, 91
Romains, Jules, 34
Romance of color, 173–189

S

Safety, color code for, 133–135, 223
  piping identification code for, 226
Sales records, 43, 55–56, 64, 209–221
  automobiles, 220–221
  blankets, 216
  carpets and rugs, 211–212
  circulars, 105
  cotton goods, 218
  envelopes, 105
  furniture, 214
  holiday novelties, 181
  home furnishings, 215–216
  kitchen ware, 218
  letterheads, 105
  linoleum, 213
  paints, 209–210
  plastics, 219–220
  reply cards, 106
  roofing, 211
  sheets, 216
  towels, 217
  trouble with, 51
Samuels, Louis, 155
Sanitary ware, color standards, 226
Schanz, F., 31
School Bus Chrome, standard for, 226
School furniture, standard colors for, 227
Schubert, 170
Schuler, Dr. Neville, 130
Schwartz, Manuel, 36
Sears, Roebuck and Co., 109
Seasons, colors of, 181
Sectional color preferences, 37–39, 60
  for hard-surface floor coverings, 60
Sedimentary rocks, color standards, 227
Senses, human, effect of color on, 35–36, 174

Settle, Captain, 140
Sexual cycles, effect of color on, 164
Shariff, 149
Shipley, A. E, 149
Signal glasses, color standards for, 227
Silver fox, control of color in, 154
Smallpox, effect of light on, 161
Social significance of color, 4, 59, 61
Soil colors, standards for identification of, 227
Soilax, 145
South (U.S.), color preferences, 60
Southall, James, 102
Southwest (U.S.), color preferences, 60
Space advertising, 108–115
  case histories, 111–114
  color technique, 114–115
  revenue for color in magazines, 109
  Starch reports, 108, 109
Spencer, D. A., 25, 169
Spencer, D. E, 79, 82
Spoehr, Dr. H. A., 153
Staples, R, 21
Starch, Daniel, 108
Starch reports, 108–110
Stein, 163
Styling, 3–5, 11–13, 49–54
  and advertising, different approaches required for, 42–43, 88–89
  economics of, 61–62
  for high-fashion markets, 3–4, 72
  for mass markets, 4–5, 11–13
  methods used by manufacturers and stores, 50–54
Sunglasses, 93
Sunlight, effect on color preference, 37–39
Symbolism, 177–182
  Freemasonry, 179
  heraldry, 179–180
  Hopi, 9
  Indians, American, 8–9
  Navahos, 8–9
  quarters of earth, 8, 9
  Roman Catholic Church, 177–178
  stock shows, 180
Synesthesia, 170–171

T

Tattooing, 9
Taves, Ernest, 157
Taylor, Helen D., 37, 71
Temperature, control of through color, 141–143
  in radiators, 142
  in storage tanks, 141
Temple of Nebuchadnezzar, 9
Textile Color Card Association, 72, 195, 227
Time estimation under influence of color, 165–166
Titchener, 194
Tomatoes, use of color in grading, 152–153
Tone, most neutral of color forms, 85
Tonus of human body, affected by color, 35, 163
Traffic devices, standards for, 227
Trends, color, 12, 57–59
  causes of, 46–47
Tumors, effect of color on growth of, 165
Twilight vision, 91

U

Ultraviolet light, 143–147
  use of color to overcome effects of, 144
Universities and colleges, color symbolism of major faculties, 180
University of California, 154
Upanishads, Hindu, 9
U.S. Agricultural Marketing Service, 199
U.S. Army Color Card, 227
U.S. Army general paint specifications, 228
U.S. Army-Navy Aircraft Standards, 228
U.S. Bureau of Mines, 141
U.S. Department of Agriculture, 144
U.S. Flag, color standards, 224
U.S. National Bureau of Standards, 142, 199, 202, 204, 222–226

U.S. Pharmacopoeia, 202
U.S. Public Health Service, 129
U.S. ribbons for service medals, color standards, 228

V

Vanderplank, 164
Victorian style, 183
Violet light, to detect mercury vapor, 154
Visibility of colors, 93, 95
Vision, 91–102
color, 192–194
of birds, 33
of fish, 33
of insects, 30–31
of lower organisms, 30
of monkeys, 33
in twilight, 92
of vertebrate animals, 30
foveal, 91
peripheral, 91
range of sensitivity to brightness, 92
training necessary, 102
twilight, 91
Visual acuity, 92–95
Visual illusions, 98–99
afterimage, 98
Benham's disk, 90, 91
brightness contrast, 99
color contrast, 99
Von Kries, 194
Von Ries, 162
Vosburg, Frederick G., 31

W

Wagner, 170
Wallis, W. Allen, 97
Walls, G. L., 33, 89, 97, 100
Walton, William E., 25
War, use of color in, 148–149
Wardrobe, average number of colors in, 49
Warmth and coolness in color, 169–170
Wave lengths, measurement of, 192
Wedgwood, Josiah, 182
Williamsburg, 183
Withrow, Dr. R. B., 32
Women's fashions, 71–73
Woodson, T. T., 154
Wounds, healing accelerated through color, 165
Wrenn Paper Company, 96

X

X-rays, sensitivity of insects to, 30

Y

Yellow, color therapy of, 162
effect on the insane, 167
glorification of, 82
meaning in character analysis, 187
in safety practice, 134
suggests form of triangle, 174
Yellowing of human eye with age, 40
Yogo, Dr., 164

Z

Zodiac, 9

Milton Keynes UK
Ingram Content Group UK Ltd.
UKHW022003140923
428719UK00005B/121

9 781016 004893